IEE ENERGY SERIES 5

Series Editors: L. Divone
Professor D. T. Swift-Hook

TIDAL
POWER

Other volumes in this series:

TIDAL POWER

A C Baker

Peter Peregrinus Ltd. on behalf of the Institution of Electrical Engineers

Published by: Peter Peregrinus Ltd., London, United Kingdom

Peter Peregrinus Ltd.,
Michael Faraday House,
Six Hills Way, Stevenage,
Herts. SG1 2AY, United Kingdom

British Library Cataloguing in Publication Data
Baker, A. C.
  Tidal power.
  1. Tidal power
  I. Title
  621.31'2134

ISBN 0 86341 189 4

Printed in England by The Redwood Press, Wiltshire

Contents

Acknowledgments

In 1978, the London-based firm of consulting engineers Binnie & Partners was appointed by the Department of Energy to be the lead consultants for a major study of tidal power from the Severn estuary. The author, having had some experience in the design of tidal defence structures and having been involved in a major feasibility study of water storage reservoirs offshore in the Wash, was made Project Engineer for this study. So began a hectic and most interesting period which culminated in the report of the Severn Barrage Committee, Energy Paper 46, published in 1981. During this period a large number of organisations and individuals were involved, bringing their experience and expertise to bear to help solve the simple questions: where should a Severn barrage be located, how should it be operated, how much electricity would it generate, how much would the electricity cost and would the environmental effects be acceptable?

Since 1981, the lead in studies of the Severn barrage has been taken up by the Severn Tidal Power Group, which published an interim report in 1986 and then a report on their extensive investigations in late 1989. The manuscript for this book was prepared in 1989 and therefore the results of the Severn Tidal Power Group's work have not been included. However, the Group's report and its supporting documents will be a valuable reference for the interested reader.

Meanwhile, studies funded by the Department of Energy have continued of a wide range of aspects of tidal power, both generic and site-specific. The author has been involved in some of these. One interesting result has been the identification of a few small estuaries in the UK which appear to be suitable for tidal power barrages of a few tens of megawatts capacity and which may be able to generate electricity at a unit rate which is economic. Two such estuaries have been selected for detailed study, the Conwy estuary in North Wales, and the Wyre estuary near Blackpool.

Outside the UK, the main areas of interest have been France, where the 240 MW Rance barrage has continued to provide useful and encouraging operational experience, and the Bay of Fundy, in Nova Scotia, where a prototype large Straflo turbine has been installed and is working well. There is, or has recently been, activity in several other countries, notably South Korea, where I have worked, India, Brazil, China, Australia and the USSR.

This book attempts to pull together in a reasonably digestible form the work that has been done on tidal power over the last 60 years. As a result, I have had to rely heavily on work done by others and published elsewhere. This, and the help of many individuals in carrying out studies over the last twelve years is acknowledged with grateful thanks. Special mention is due to Prof. Eric Wilson of Salford University, to Dr David Keiller who has been responsible for most of the computer modelling of tidal power schemes carried out by Binnie & Partners, and to Dr Roger Price of the Energy Technology Support Unit at Harwell, who has been the Project Officer for most of the work carried out by Binnie & Partners for the Department of Energy. Finally, my thanks to Rosalind Hann who patiently steered this book through the process of publishing.

Clive Baker 1991

Chapter 1

Tides

1.1 Introduction

The understanding and prediction of the tides are subjects about which many books and learned papers have been written. Good examples of the former are Dronkers (1964(1)) and Pugh (1987(15)). The aim of this chapter is to provide a summary of the aspects of tides which are relevant to tidal power barrages. First, a brief summary is given of the origin of the tides.

1.2 The origin of the tides

The earth rotating on its axis generates large centrifugal forces. These result in the diameter of the earth at the equator being about 21 km larger than that through the poles, and also increases the depth of the seas at the equator relative to that at the poles. This variation in water depth does not vary with time, except that the speed of rotation of the earth is decreasing very slowly, and so can be considered constant and does not result in tides.

The tides are generated by the rotation of the earth within the gravitational fields of the moon and the sun. The loss of energy as heat due to friction caused by the tides appears to have helped the time of rotation of the earth to increase from about 8 hours when the earth was formed to its present value of 24 hours. Thus the ultimate source of energy for the tides may be the rotation of the earth. The planets also have effects on the tides which, because of the distances involved, are minute and so can be ignored. The starting point is Newton's law of gravity; namely that a body exerts an attracting force on another body, the strength of this force being proportional to the product of the masses of the two bodies and inversely proportional to the square of the distance between their centres of mass.

The earth has a mass of about 6×10^{21} tonnes. The masses of the sun, moon and earth (S, M and E, respectively) are in the following ratios

$$M/E = 1/81 \cdot 3 = 0 \cdot 0123$$

$$S/E = 333\,000$$

The mean distances of the sun (D_s) and moon (D) from the earth are about 150 000 000 km and 384 000 km respectively, a ratio of about 391:1.

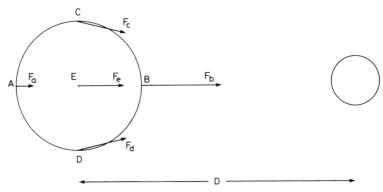

earth (mass E, radius r)   moon (mass M)

Fig. 1.1   The vector field showing the moon's gravitational force on the earth at locations in the equatorial plane

From this it follows that the gravitational pulls of the moon and sun on the earth are in the ratios

$$1 \times 0{\cdot}0123/1^2 : 1 \times 333\,000/391^2$$

i.e. about 1:177.

Thus the sun should be by far the dominant force in the development of tides. However, from earliest times people have been aware that the moon is the principal influence on the time and height of tides. Consequently some factor other than straight gravitional attraction is at work.

For a simple explanation of the development of tides, the earth is assumed to be covered in a uniform depth of water. Since about 75% of the earth's surface is covered by water, this is not too far from the truth, although the presence of the continental land masses complicates the actual development of the tides greatly.

Consider the effects on the earth of the gravity field of the moon only (Fig. 1.1). The magnitudes of the force vectors are given by Newton's law of gravitation and all vectors point to the centre of the moon. Point B on the side of the earth nearest the moon will be subject to a greater force than that at the centre of the earth, while point A will be subject to a smaller force. If the 'average' force at the centre is deducted from the force on each point, then the forces shown in Fig. 1.2 are left. These are the forces which generate tides.

The gravitational constant G, which determines the attraction between unit masses at unit distance, is $6{\cdot}67 \times 10^{-8}$ dyne cm^2/gm^2 (or $6{\cdot}67 \times 10^{-11}$ N m^2/kg^2). Thus the acceleration due to gravity at the earth's surfaces g, is given by $g = GE/r^2$, where r is the mean radius of the earth (6370 km), and is about 9.81 m/s^2.

Considering now the tide-generating force T_b in Fig. 1.2, this will be given by

$$T_b = \frac{GM}{(D-r)^2} - \frac{GM}{D^2} \simeq \frac{GM}{D^2}\left(1 + \frac{2r}{D}\right) - \frac{GM}{D^2} \simeq \frac{2GMr}{D^3}$$

Similarly, the force T_a will be given by

$$T_a = \frac{GM}{(D+r)^2} - \frac{GM}{D^2} \simeq \frac{GM}{D^2}\left(1 - \frac{2r}{D}\right) - \frac{GM}{D^2} \simeq -\frac{2GMr}{D^3}$$

and T_c and T_d will be given by

$$T_c = T_d = F_c \cos\theta = \frac{GM}{D^2 + r^2} \times \frac{r}{(D^2 + r^2)^{1/2}} = \frac{GMr}{(D^2 + r^2)^{3/2}} \simeq \frac{GMr}{D^3}$$

Assuming the distance of the earth from the moon to be constant, the tide-generating forces discussed above are constant with respect to the angle subtended between each point, the moon and the centre of the earth. Thus, if the earth were not rotating about its axis relative to the moon, the effect of these forces would be a slight permanent distortion of the water surface, which would be difficult to detect, with the forces on the water particles in equilibrium (and the surface at each point at right angles to the resultant force). A similar distortion is caused by the sun. The magnitude of this relative to that caused by the moon is given by

$$\frac{SD^3}{MDs^3}$$

Substituting the values given earlier gives the sun's effect as 46% of that of the moon. This helps explain why the moon dominates the tides in spite of its gravitational force being much smaller than that of the sun; in essence, the ratio of the earth's radius to the distance of the moon from the earth is much greater

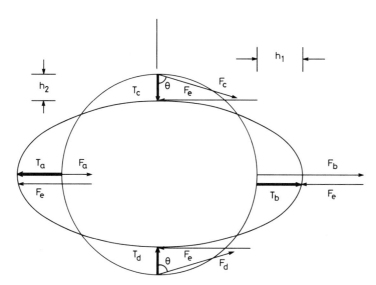

Fig. 1.2 The tidal generating force field produced by the vector addition of $(-F_e)$ and the peripheral forces

than its ratio to the distance of the sun from the earth, and it is the difference in distance of opposite sides of the earth which causes differences in gravitational fields.

To estimate the 'equilibrium' tidal range due to the moon, the height h, is calculated at which total gravitational forces balance. With g (= force per unit mass) given by $g = GE/r^2$, the value g_h at some height h above mean sea level will be

$$g_h = \frac{GE}{(r+h)^2} = \frac{GE}{r^2}\left(1 - \frac{2h}{r}\right)$$

As shown in Fig. 1.3, the ocean surface will rise until forces are in equilibrium. For point A, this gives

$$g - T_a = gh$$

$$g = gh + T_a$$

$$\frac{GE}{r^2} = \frac{GE}{r^2}\left(1 - \frac{2h}{r}\right) - \frac{2GMr}{D^3}$$

Solving for h,

$$h = \frac{M}{E}\frac{r^4}{D^3}$$

and for point C, this gives

$$h = -\frac{M}{E}\frac{r^4}{2D^3}$$

Thus the total range is $\dfrac{3Mr^4}{2ED^3}$ which is approximately 560 mm.

A more precise calculation is given by Dronkers (1964(1)), who shows that the mean tide would be 534 mm, with the difference on opposite sides being 24 mm.

These calculations assume that the earth is covered uniformly with water, without any protruding land masses. The presence of the continents prevents this assumption being correct, but it broadly applies to the Atlantic, Pacific and Indian oceans. Tides do occur in other seas such as the Mediterranean and the Red Sea, and are described in detail by Pugh (1987(15)), but are too small to be of interest for tidal power.

Fig. 1.2 shows the general form of the forces at the earth's surface which would cause this permanent distortion, in addition to the permanent distortion caused by centrifugal forces, if the earth were not rotating. With the earth rotating, any given point will become subject to cyclic horizontal forces which are zero when the moon is overhead, when the moon is diametrically 'underneath' and when the moon is on the horizon, and reach maxima when the elevation of the moon is about 45°. These are illustrated in Fig. 1.4.

The earth rotates within this force field once every 24 hours. In addition, the moon orbits the earth every 27·3 days in the same direction as the rotation of the earth, and both are rotating in the same direction as the earth's orbit around the sun. As a result, the time between successive full moons is 29·5 days. This means that the time between successive transits over a selected point on the surface of the earth is about 24 hours and 50 minutes. This is the interval between successive high waters on the side nearest the moon, and also on the side away from the moon. Consequently the time between high tides at any one place is normally about 12 hours and 25 minutes.

The next question concerns the conversion of an open-sea range of about 550 mm to ranges of 10 m or more, as occurs in the Bay of Fundy, the Severn estuary and so forth. This process results from a combination of two factors. Firstly, the deep-water tidal wave increases in height and slows down as it enters shallow water. This process starts as the wave crosses the continental shelf. Secondly, two stages of resonance are involved. Resonance is the result of a body being subjected to a small, regular impulse at a frequency which is the same as, or a multiple of, the natural frequency of vibration or oscillation of that body. A clock's escapement mechanism is a simple example. In the oceans, the water particles are each subject to a pull and a push in every 12 hours 25 minutes, as shown in Fig. 1.4. The speed (V) of a tidal wave whose height is small in relation to the depth of water (d), and whose wavelength is long in relation to the depth, is defined by the relationship

$$V = (g\mathrm{d})^{0.5}$$

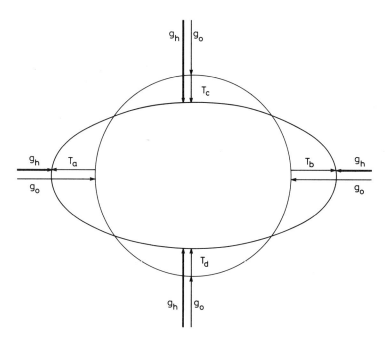

Fig. 1.3 Equilibrium of the oceanic tidal distortion – the value of g_h is the vector addition of g_0 and the tidal force

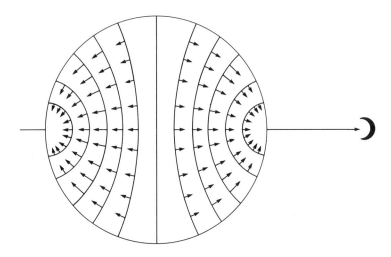

Fig. 1.4   Moon's tractive forces on the earth's surface

and the length of the wave (L) between successive crests will be given by

$L = Vt$ where t is the wave's period.

In the open ocean, with a water depth of say 3000 m, a tidal wave will travel at about 170 m/s. In the north Atlantic, the distance between America and Europe is about 4000 km. If the ocean had vertical sides so that the tidal wave reflected efficiently, then resonance would occur if the wavelength was twice the width of the ocean (Fig. 1.5). Thus a wavelength of about 8000 m with a period of 8 000 000/170 × 3600 = 13 hours should result in resonance. This is close to the frequency of the tide-generating force from the moon and results in a tide range greater than the equilibrium range. As the deep-water tidal wave crosses the edge of the continental shelf, its frequency remains the same and its height increases, but its length and therefore its speed decrease. This results in a tidal range of about 3 m on each side of the north Atlantic.

The Pacific is about four times wider than the Atlantic, and so its natural frequency is closer to 24 hours, but since its width varies greatly, the pattern of tides is more complex and there are fewer locations with large tidal ranges.

The second stage of resonance causes the tidal range at the edge of the ocean to increase dramatically in certain estuaries, and is caused when the natural frequency of the 'driving' tide at the mouth of the estuary is close to, or a low multiple of, the natural frequency of tidal propagation up the estuary. An important factor now is the decreasing speed of the tidal wave as the sea becomes shallow. In the western approaches to Europe, the speed of the wave decreases to around 50 m/s, and then, in the Severn estuary where the average depth of water is about 25 m, to about 15 m/s.

Thus the Celtic Sea, Bristol Channel and Severn estuary, with a total length of about 600 km, have a natural frequency of about 6 hours, and are resonating as a quarter wave due to a tidal impulse with a frequency of about twice this.

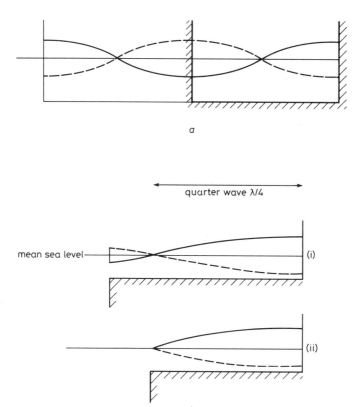

quarter wave λ/4

mean sea level

(i)

(ii)

b

Fig. 1.5 Types of resonance
a Resonance with wavelength = 2 × basin width
b Quarter wave resonance of basin driven by tidal level changes at the open end

Fong & Heaps (1978(3)) discuss this in detail. The distance from Land's End to Dover, along the English Channel, is about 500 km and the average depth is about 70 m. Therefore this 'estuary' has a half-wave oscillation, with high and low waters occurring at Dover and off Cornwall respectively at the same time. Coriolis force, discussed later, results in the tidal range on the south side of the Channel, for example at St. Malo, being much higher than on the north.

An estuary may have a length which is longer than the ideal to achieve resonance. It is therefore theoretically possible that the introduction of a tidal barrage, by attenuating an estuary, could increase the tidal range to seaward of the barrage. Early work with computer models of the west coast of the UK indicated that an impermeable barrier located near the proposed site of the Severn barrage would increase the amplitude of the mean tide ($M2$) constituent slightly (Ref. 1968(1)). However, when a working barrage was substituted, the damping effect this would have (and the energy extracted) on the tide was then predicted to reduce the tidal range to seaward.

1.3 Diurnal tides

The discussion so far has concentrated on semi-diurnal tides. These are the most important as regards possible tidal power developments, but are subject to variations arising from the axis of rotation of the earth being inclined to the planes of orbit of the moon around the earth and the earth around the sun. The lunar equilibrium tide is aligned with the orbit of the moon, and this is inclined at between 18·3° north and 28·6° south of the equator. Similarly, the sun's declination varies between 23·5° north and south of the equator, these angles of latitude defining the tropics. During each rotation of the earth, a point on the earth's surface will pass through different parts of the equilibrium tide envelope and therefore experience a diurnal variation in tide levels. This is illustrated in Fig. 1.6. Examples of tide curves showing this effect are given in Chapter 12.

1.4 Coriolis force

North of the equator, a current which moves northwards results in that body of water moving nearer the earth's axis of rotation. Consequently, it moves to a location where less angular momentum (in the east–west direction) is required to maintain its relative east–west position. Since the earth is rotating eastwards, the body of water has to dissipate energy in this direction of movement and therefore moves to the east. The opposite happens to a current moving southwards, and the mechanism is mirrored south of the equator. This force

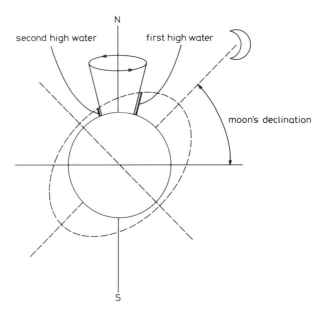

Fig. 1.6   Effect of moon's declination on diurnal tide

causing currents to deflect is called the Coriolis force, and is proportionately more important in higher latitudes and faster currents.

Coriolis force is important for tidal power because it causes tide ranges to be distorted. For example, tidal ranges in the Channel are higher on the French coast than on the south coast of England (Fig. 1.7). Similarly, the tidal range on the east side of the Irish Sea, particularly in Liverpool Bay, is much higher than on the east coast of Ireland.

1.5 Variation of tidal range

The principal and most apparent variations in tidal range are caused by the relative positions of the sun and moon. When these are in line, i.e. at full moon and new moon, the highest, spring, tides occur. When at right angles, the first and third quarters of the moon's cycle, then neap tides occur. Since the gravitational effect of the sun in generating tides is 46% of that of the moon, spring, mean and neap tide ranges will in theory be in the ratio of $(1 + 0.46):1:(1 - 0.46)$, i.e. an overall range of about 2.3:1. In practice, this ratio is altered by local variations in the tidal constituents.

1.6 Tidal constituents

Strictly, mean spring and mean neap tides in deep water have been considered so far, comprising tides with amplitudes of $(M2 + S2)$ and $(M2 - S2)$ respectively, $M2$ being the moon's principal twice-daily or semi-diurnal constituent, and $S2$ the sun's principal semi-diurnal constituent. In this context, the amplitude is half the range. Because the motions of the sun and moon in their apparent orbits around the earth are highly complex, the pattern of tides at any location is also highly complex, varying from day to day, week to week and year to year. Some factors causing these variations are:

- The moon's orbit being elliptical
- The earth's orbit being elliptical
- The earth's axis of rotation being inclined to its orbit around the sun
- The moon's orbit being inclined to the earth's axis of rotation, the angle of declination also varying
- Shallow water effects
- Atmospheric effects

As an example, Fig. 1.8 shows all the tides predicted for the Wash, a large bay on the east coast of England, in one year (Ref. 1976(4)). The observed tides can be analysed and broken down into a large number of 'harmonic constituents', each of which is sinusoidal and modifies the two basic constituents $M2$ and $S2$. This work is normally done by a specialist institute of oceanography. Table 1.1 includes a list of the constituents for Flat Holm in the Severn Estuary (Ref. 1981(16)). Each constituent is a sine wave whose height at any time is defined by a modified version of the simple relationship

where $H = a \cos \omega t$

a is the amplitude (half the range)

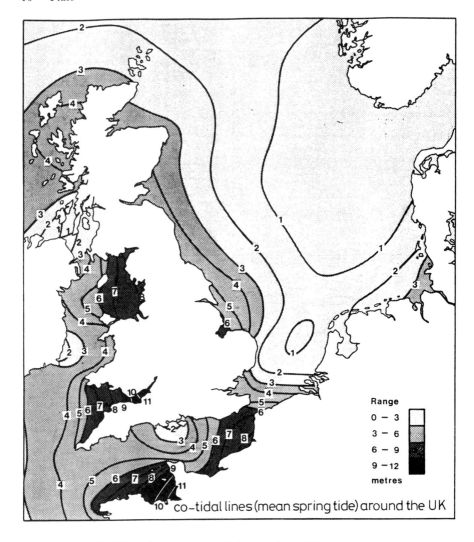

Fig. 1.7   Co-tidal lines (mean spring tide) around the UK

ω is the speed of rotation (deg/h or rad/h), and t is time elapsed since the selected starting point.

The harmonic constituents fall into bands, depending on their period. These are:

- Long period (2 weeks to 21000 years)
- Diurnal, with periods of 22 to 28 hours
- Semi-diurnal, with periods of 11·5 to 13·2 hours
- Short period, with periods of 3 to 8 hours.

Table 1.1   Tidal constituents at Flat Holm (from Ref. 81(16))

H	ω	H	ω	H	ω	H	ω
A.Z	3·713	—	$2Q_1$	0·005	63·8	OQ_2	0·039	71·9	MO_2	0·009	197·7
Sa	0·062	192·7	O'_1	0·005	138·8	MNS_2	0·097	241·8	M_3	0·043	192·4
Ssa	0·018	322·3	Q_1	0·031	328·7	$2N_2$	0·051	230·2	SO_3	0·011	161·4
Mm	0·033	347·9	ρ_1	0·005	171·6	μ_2	0·355	251·4	MK_3	0·012	252·4
MSf	0·071	42·3	O_1	0·083	3·4	N_2	0·723	174·2	SK_2	0·021	305·4
Mf	0·050	102·6	MP_1	0·005	290·2	ν_2	0·168	153·6			
			M_1	0·012	202·6	OP_2	0·058	147·5	MN_4	0·061	3·6
			X_1	0·004	321·6	M_2	3·895	190·0	M_4	0·163	31·6
			π_1	0·002	177·5	MKS_2	0·063	335·7	SN_4	0·014	58·4
			P_1	0·030	130·2	λ_2	0·106	173·0	MS_4	0·084	58·4
			S_1	0·015	65·9	L_2	0·285	168·3	MK_4	0·030	74·2
			K_1	0·063	139·5	T_2	0·073	239·6	S_4	0·027	86·2
			ψ_1	0·004	170·8	S_2	1·350	245·6	SK_4	0·012	191·4
			ϕ_1	0·004	181·7	R_2	0·016	269·9			
			θ_1	0·003	274·6	K_2	0·392	243·1	$2MN_6$	0·013	205·8
			J_1	0·004	250·8	MSN_2	0·077	45·3	M_6	0·027	224·5
			SO_1	0·006	38·2	KJ_2	0·017	176·4	MSN_6	0·011	244·0
			OO_1	0·005	322·8	$2SM_2$	0·098	68·4	$2MS_6$	0·035	281·7
									$2MK_6$	0·008	226·9
									$2SM_6$	0·015	271·3
									MSK_6	0·014	0·6

The height of the tide at any location and at any time can be calculated by adding together all the constituents, taking into account the amplitude, frequency and phase of each. This addition takes the form

$$H = \sum_{n=1}^{n} f_n h_n \cos(E_n + \omega_n T - g_n)$$

where H is tide level at the time in question

n is the number of the constituent

h_n is the constituent's mean amplitude at the site during the 18.6 year period

f_n is the constituent's phase (called astronomical argument) at time zero

ω_n is the constituent's angular speed (rad/h or deg/h)

T is time since zero (hours)

g_n is the phase lag of the constituent at the site behind the corresponding constituent at Greenwich.

When using tide tables e.g. Admiralty Tide Tables, (1988(1)), levels are given relative to chart datum, which is normally the lowest astronomical tide level, i.e. ignoring negative surges or other atmospheric effects. Thus an additional 'constituent' is listed, Z_0, which is the height of mean sea level above datum and has to be added to H when working with charts.

The speed of each constituent, ω_n, is defined by tidal theory. For $S2$, the principal solar constituent, it is exactly $30°/\text{hour}$, i.e. one rotation of the earth ($360°$) divided by 12 hours, or $30/\pi$ radians/hour. Similarly, because the sun passes over the Greenwich meridian (longitude $0°$) at noon each day, the argument of $S2$ is always 0. Although astronomers use radians when defining speeds, tides are usually defined in degrees.

The local amplitude h_n and phase lag g_n of each constituent are known as the harmonic constants and vary from site to site. The phase angle is similar to the argument but applies locally and thus takes into account the effects of the position of the point of interest in retarding the angle of the constituent.

The nodal factors are published at the tide tables; daily values for the principal constituents $M2$, $S2$, $O1$ and $K1$, and monthly values, or factors with which to multiply the nodal factor for a main constituent, for the minor constituents.

1.7 Effects of individual tidal constituents

The most important constituent is $M2$ which has a speed of about $28 \cdot 984°/\text{hour}$. The resulting tide range $2 \times M2$ will be the range of the mean tide and this governs the overall energy output of a barrage. The next most important is $S2$, which causes the principal spring/neap variation. $M2$ and $S2$ beat together every $14 \cdot 77$ days to form the typical spring/neap cycle. In shallow water, corrections have to be applied. Admiralty Tide Tables set these out in the form

$$\text{MHWS} = Z_0 + (M_2 + S_2) + F_4(M_2 + S_2)^2 \cos f_4 + F_6(M_2 + S_2)^3 \cos f_6 + M\,Sf$$

A tide of range $2 \times (M2 + S2)$ will be close to the mean spring tide and form the basis of factors such as the maximum differential head across the barrage, the maximum power from the turbines, the maximum tidal currents at the site and so forth. As the period of $14 \cdot 77$ days does not contain a complete number of cycles, the pattern does not repeat exactly every $14 \cdot 77$ days but very nearly repeats every 251 days. Since these two constituents are based on circular astronomic orbits, corrections for the elliptical orbits of the moon and the earth are made by constituents $N2$ and $K2$ respectively. $N2$ beats with $M2$ every 27.55 days and results in alternate sets of spring tides being higher and lower than average, and also affects neaps in the same way but out of phase with springs. This effect can be seen in Fig. 1.8.

The $K2$ constituent beats with $S2$ every $182 \cdot 6$ days or twice a year, and is in phase in late March and late September and out of phase in June and December. The former results in the equinoctal spring tides which are usually the highest of the year. Overall, a fairly close repeat of the beat cycles of $M2$, $S2$, $N2$ and $K2$ will occur every 2009 days or 5.5 years. This, combined with the long term variation of the declination of the moon's orbit, means that the total annual energy output of a tidal barrage will vary significantly over an 18.6 year cycle and care has to be taken when selecting a year's tides on which to base

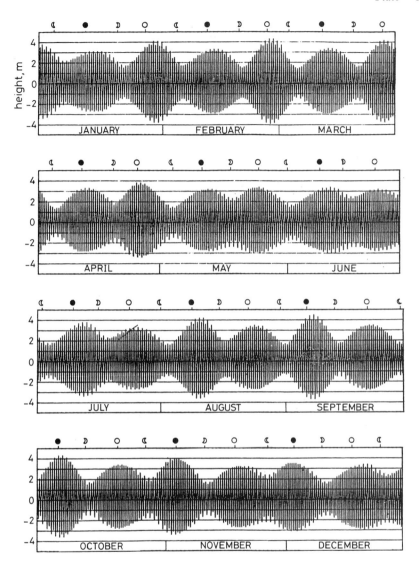

Fig. 1.8 Predicted tides for The Wash, UK, 1975

energy calculations. As an example, Fig. 1.9 compares the tides at Avonmouth in the Severn estuary for the years 1969 and 1978. The much greater number of spring tides in 1978 is apparent. The year 1974 was selected for energy calculations in Refs. 1980(5) and 1981(1) as being close to a year of mean tides. 1983 and 1992 will similarly be years close to the mean.

The principal diurnal constituents are $K1$ and $O1$ and are generally more important around the shores of the Pacific ocean. Their effect is to cause

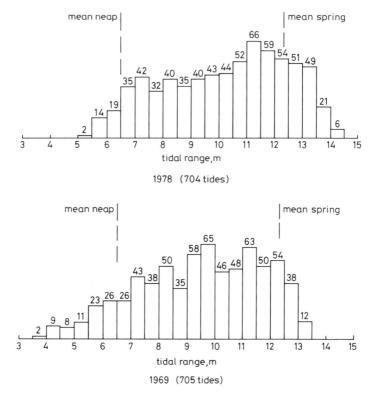

Fig. 1.9   Annual tide range distributions for Avonmouth

alternate high-tide levels to be higher and lower than each other and, similarly, alternate low-tide levels to vary. Examples are given in Chapter 12.

Short-period constituents such as MS_4 have the effect of distorting the sine of wave of the main constituents as the tide runs into a shallowing estuary. Thus the tide floods more quickly and ebbs more slowly. In certain estuaries, the distortion becomes so great that, during high spring tides, the flood tide in the uppermost part of the estuary rises instantaneously and therefore forms a bore. Famous examples of estuaries with bores are the Severn in England (Fig. 1.10) and the Hooghly near Calcutta. Fig. 1.11 shows how the tide running up the Severn progressively changes in shape in this manner, culminating in a flood tide rise in the upper estuary which is almost vertical, i.e. a bore (Ref. 1981(1)).

The discussions above have been concerned with the calculation of the height at a given location for a selected time. This will be appropriate when considering the overall performance of a tidal barrage throughout a year, including, for example, the contributions of the barrage output to peak electricity demands. For the simple tidal energy calculations which define the output of a barrage over an average spring/neap cycle, actual times of high water are not needed, so phase angles and arguments can be ignored. Instead,

Fig. 1.10 The Severn bore
(Photograph: Helen Sidebotham)

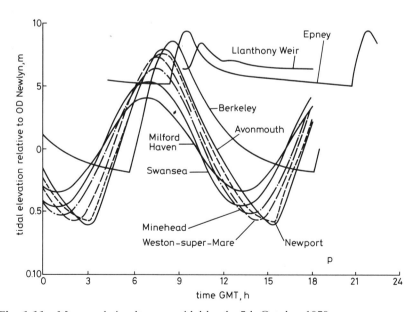

Fig. 1.11 Measured simultaneous tidal levels: 7th October 1979

Table 1.2   Typical relationship between time of high water and tidal range, for Severn barrage

Tidal range (m) between	4·4 5·2	5·3 6·2	6·3 7·2	7·3 8·2	8·3 9·2	9·3 10·2	10·3 11·2	11·3 12·2	12·3 12·7	Total
Time of high water between										
0000 0059	6	5	3	1	2					17
0100 0159	6	8	5	3						22
0200 0259	2	5	9	3						19
0300 0359		3	5	7	4					19
0400 0459			3	4	15	5				27
0500 0559					8	11	10	1		30
0600 0659					1	10	12	13		36
0700 0759						8	11	16	4	39
0800 0859					2	11	12	13	3	41
0900 0959					5	10	13	10	3	41
1000 1059		2	7	9	6	7				31
1100 1159	1	9	7	7	5					29
1200 1259	7	10	5	3						25
1300 1359	3	6	4	3						16
1400 1459	1	6	7	2						16
1500 1559		3	5	9	5					22
1600 1659			3	6	9	9				27
1700 1759				1	7	12	10	2		32
1800 1859					1	9	11	12	1	34
1900 1959						9	10	15	5	39
2000 2059					2	10	12	15	1	40
2100 2159				5	12	12	11	2		42
2200 2259		2	10	12	6	5	1			36
2300 2359	1	9	8	3	2	1				24
Totals	27	68	81	83	97	132	110	42	14	

the cycle can be defined simply by the expression

$$h = H_{M2} \cos(0·5058t) + H_{S2} \cos(0·5236t)$$

This starts at high water of a mean spring tide. In practice, at any given location, the time of high water and tidal range for each tide in a year near the proposed site of the Severn barrage have been assembled as a histogram. This shows how high water of spring tides is concentrated at around 8 a.m. and 8 p.m. This can be important when assessing the value of the electricity produced by a particular barrage. In Liverpool Bay, high water occurs about 5 hours later than in the Severn, so that spring high waters are concentrated around midday and midnight.

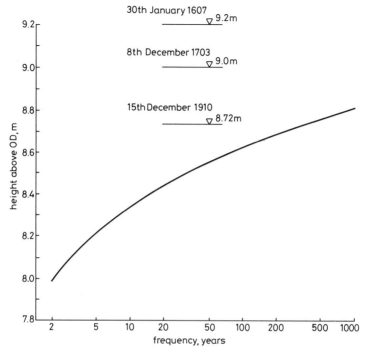

Fig. 1.12   Frequency of abnormally high tides at Avonmouth

1.8 Surge tides

The actual height and time of high water can and do vary significantly from those predicted in tide tables. This variability is caused by meteorological factors, namely the atmospheric pressure and wind strength and direction. A millibar (one thousandth of an atmosphere) change will change the sea level by about 10 mm. Thus a deep depression of say 970 mb will raise the sea level by 300 mm. Strong winds, especially those associated with deep depressions, can generate a slope or 'set up' the surface of the sea. Together, these effects can increase the level of the sea at the coast substantially. These short-term changes are called 'surges'. They can be positive or negative and can arrive at the coast at any state of the tide, but clearly the most noticeable and important are those arriving around the time of high water which increase the height of the tide. Then overtopping and/or breaching of sea defences can result, with consequent loss of life and damage to property. Fig. 1.12 (Ref. 1981(1)) shows the predicted frequency of exceptional high waters in the Severn estuary. Their effect on the energy output of a barrage will be short-term and thus unimportant, although prediction of such surges a few hours ahead will be important in the operation of a barrage. Their main interest as regards tidal power lies in their effects on the design of the barrage and on the possibility of the barrage being used to exclude these surges and so prevent flooding in the enclosed basin, this also affecting the design of the sluice gates.

1.9 Tidal data

If a tidal power site is in a location not covered by tide tables, which is often the case if the estuary is not used by shipping, approximate information can be obtained by reference to the nearest standard or secondary port in the tide tables, and interpolating or extrapolating accordingly. If the tidal range is varying rapidly along that part of the coast, then there is a risk of inaccuracy.

In the absence of adequate data, the tides will have to be measured at the proposed site, preferably by a continuously recording automatic tide gauge. The constituents can then be extracted by computer. Oceanographic institutes have the facilities to do this. The period over which readings have to be taken for the main constituents to be identified is surprisingly short, as long as the data are accurate and there have not been unusual atmospheric conditions. Thus adequate accuracy for a feasibility study will be reached with readings for a single spring tide and a single neap tide, recorded simultaneously at a number of locations along the length of the estuary so that the dynamics of the propagation of the tide can be investigated. One month's readings will enable the main constituents to be extracted. For precise definition, a year's readings are needed.

Chapter 2

The operation of a tidal power barrage

2.1 Principal components

A tidal barrage is designed to extract energy from the rise and fall of the tides and is relatively simple in concept. Thus a tidal barrage has only four main components:

- *Turbines*, located in water passages which are designed to convert the potential energy of the difference in water levels across the barrage into kinetic energy in the form of fast-moving water. This kinetic energy is then converted into rotational energy by the blades of the turbines and then into electricity by generators driven by the turbines.

- *Openings* fitted with control gates, called sluices, which are designed to pass large flows under modest differential heads. These have a dual role. During construction, they allow the tides to continue to flow into and out of the basin behind the barrage with relatively little obstruction and thus enable the last parts of the barrage to be built without undue difficulty. Once the barrage is in operation, they refill the basin (or empty it) ready for the next power generation period.

- *Locks* or similar apparatus, to enable ships or boats to pass safely across the barrage after it is complete, and to pass safely through a part complete barrage where the remaining openings would have relatively fast-flowing water and/or construction activities in progress.

- *Embankments*, or else simple concrete caissons, which fill the remaining gaps across the estuary. These have to be reasonably 'opaque' to water flow so that water and energy are not wasted. They also provide a route to the working parts of the barrage for operation and maintenance staff and equipment, and a safe route for power cables from the barrage to the shore.

With these basic components, a tidal power barrage can be designed to operate in various 'modes'. These are described and discussed in the following sections.

2.2 Ebb generation

This mode of operation is so named because the direction of flow during power generation is the same as the ebb tide, i.e. towards the sea. Fig. 2.1 illustrates the extremes of possible range in water levels in the basin. If a single turbine is

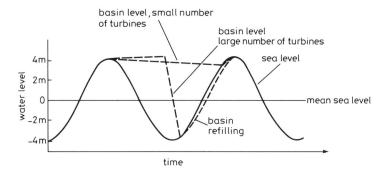

Fig. 2.1   Extreme operating cycles

installed in a barrage which encloses a basin large enough to supply a number of similar turbines satisfactorily, the basin level will hardly change between one high tide and the next. The energy output per unit installed capacity (MWh/kW) for that turbine will be large, because it will be operating at an average head which is a large proportion of the tidal range, and it will be operating for almost the full tidal cycle. On the other hand, a very large number of turbines could be installed, so that the basin could be drained rapidly at about the time of low water, as also shown in Fig. 2.1. This would result in the maximum extraction of energy from the water, but the energy would be produced in such a short time that the instantaneous power produced would be large and difficult to absorb. Because the turbines would be operating for a very short time each tidal cycle, the energy output per unit installed capacity would be small and the cost of the energy high. In addition, the rapid release of large quantities of water from the basin would reduce the effective head across the turbines, because there would be a slope on the water surface down towards the barrage on the basin side, and a steeper slope away from the barrage on the seaward side because the flow would be in a shallower and narrower channel. Very rapid releases of water would also cause various problems with the environment and would only benefit white-water canooists.

A typical, near-optimum, operating cycle is shown in Fig. 2.2, which shows the operation of a barrage over a 14-tide cycle, from mean springs to mean neaps. Also shown is the energy output during each tide, demonstrating clearly the progressive reduction by a factor of about six from springs to neaps. Three other points come out. Firstly, for maximum energy production, the turbines do not start until well after the sea has started to ebb, typically about three hours. Secondly, the maximum water level in the basin is only slightly below high water on the seaward side of the barrage. This is important because every drop of water let into the basin represents energy to be won. Thirdly, the new minimum water level in the basin is close to normal mean sea level, reached when the head across the turbines is the minimum at which they can operate properly. When the turbines close down, there is a short period when the basin water level remains constant until the next flood tide rises above basin water level and the basin can be refilled.

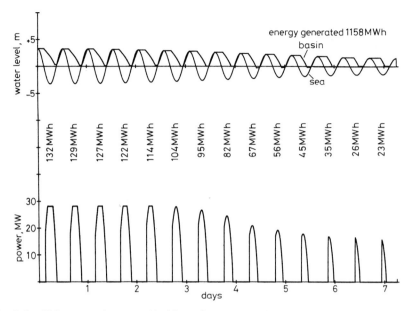

Fig. 2.2　Ebb generation over 14 tide spring-neap cycle

Fig. 2.3 is of some historical interest. It shows the operating cycle, somewhat simplified for case of calculation for an 'average' tide recommended in Ref. 1923(1), a very early study of tidal power in the Severn, and bears close resemblance to the cycle shown in Fig. 2.2. In the paper referred to, ebb generation, flood generation (discussed next) and two-way generation are compared and the conclusion drawn that ebb generation would be the best mode of operation. This conclusion was the same as that reached by the Severn Barrage Committee in 1981 (Ref. 1981(1)).

In the operating cycle shown in Fig. 2.2, the flow through the turbines is defined as a single curve relating turbine flow and power to the head across the barrage. This is discussed in Chapter 3. Similarly, the flows through the sluices and the turbines acting as orifices during the refilling of the basin are defined in the form (Section 5.3):

$$Q = AC_d(2gH)^{0.5}$$

The principal variable remaining is the time at which the turbines are started, i.e. the starting head. The optimum head is normally that which results in the maximum energy output for that tide. Fig. 2.4 shows the results of sensitivity tests of the effects of early or late starting on energy output for different tidal ranges (Ref. 1986(13)). This demonstrates that starting head is less critical, in percentage terms, for spring tides than neap tides, partly because a tidal barrage is likely to have an installed generator capacity which is less than that needed to generate the power available during relatively few large spring tides, so that the turbines have to be throttled for a while in any case. This effect can be seen as the flat-topped energy output curves in Fig. 2.2.

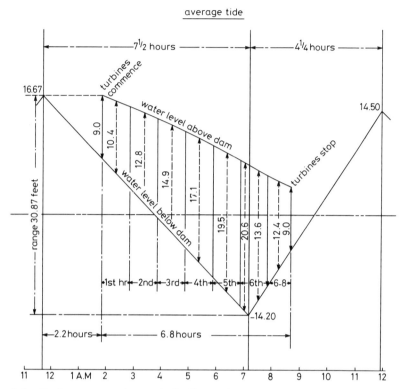

Fig. 2.3   Optimum operating cycle for Severn barrage as assessed in 1923

Optimising the turbines' starting head for maximum energy is the usual starting point. However, the value of the energy generated and supplied is not normally uniform throughout the day or even the week; energy generated during times of peak demand can have a value much higher than energy produced in the early hours of a weekend morning. There will clearly be some scope for advancing or retarding the time that generation starts on a particular tide in order to take advantage of any changes in energy value. Here the predictability of the time of high water, and hence the time of power production, is a great advantage for tidal power over, for example, wind energy.

Figs. 2.2 and 2.3, like Fig. 2.1, are based on so-called 'flat-estuary' model results, which assume that the tide on the seaward side of the barrage is unchanged in shape, and that the water level in the basin is flat. Thus a volume V entering or leaving the basin is assumed to change the basin level by V/A, A being the basin area at that level. This method of calculation can provide reasonably accurate results for a barrage in a short estuary where the water surface remains reasonably flat during the ebb and flow of the tide. Much more accurate results, particularly for sites in long estuaries, are obtained if the dynamics of water flows are included in the calculations. This subject is discussed in more detail in Chapter 9.

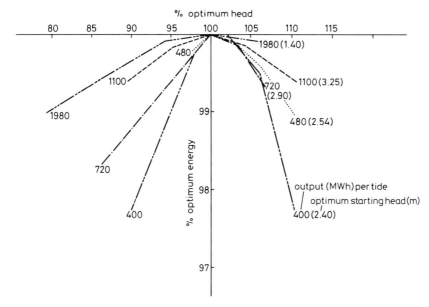

Fig. 2.4 Effect of generator start head on energy output

Fig. 2.5 shows the water levels on either side of the Severn barrage for a mean spring tide, as predicted by a dynamic computer model (Ref. 1980(5)). Starting at point A, the sequence of events is as follows.

The basin behind the barrage is filled by water flowing into the estuary during the flood tide through the open sluices and also the turbines freewheeling in reverse. The minimum obstruction is presented to the rising tide. Thus the provision of adequate sluice area is important.

At B, shortly after high tide, the basin level, which has been lagging behind the rise in sea level due to losses across the barrage, will equal the level of the ebbing sea. This is the moment for the sluice gates and the turbine flow control gates to be closed. In practice, neither type can be closed instantaneously, so closure has to start slightly before the optimum time. If the enclosed basin is long, such as the Severn estuary, then the water level immediately behind the barrage could continue to rise as the tidal wave runs back from the top of the estuary at the start of the ebb. If the estuary is relatively short, this dynamic effect will not be significant.

At C, some time after high tide, the turbines are started. The precise time for starting a particular turbine will depend on several factors, including:

• Whether the rate of starting the turbines would pose any problems for the grid system in reducing the output of thermal or hydroelectric stations to allow the output of the barrage to be absorbed efficiently and to the best advantage in terms of fuel saved. If this were the case, then the turbines would be started in sequence, beginning before the time which would result in the maximum energy production.

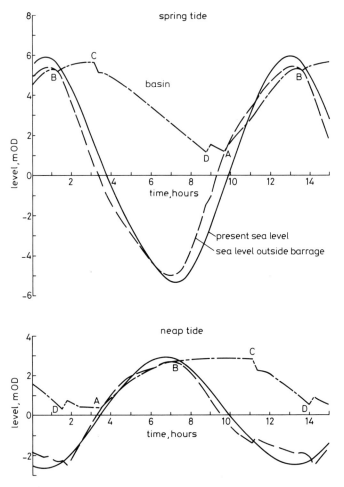

Fig. 2.5   Water levels on both sides of a barrage at Flat Holm

● Whether there would be any problems in the enclosed basin or in the estuary to seaward of the barrage from sudden changes in water level caused by a rapid opening up of the turbines. Again, if there were, then the turbines would have to be started in sequence.

● Whether there were any changes in tariff, or the value of the electricity produced by the barrage, during the generating period which would make it beneficial to concentrate electricity production towards the beginning or the end of the period. The former would result in the start time being brought forward, the latter in its being delayed.

● Variation in the actual water levels either side of the turbines along the length of the barrage. This effect would only be noticeable along a long barrage and could be caused either by asymmetry in the geometry of the estuary, so that the ebb tide starts earlier on one side of the estuary than on the other, or by strong winds causing the water surface to 'set up' along the barrage.

The turbines having started, operation is largely governed by the fact that the water stored behind the barrage will be 'wasted' if it is not used before the next flood tide rises to a level which results in a head across the barrage which is the minimum that the turbines can operate at, typically a level difference of 1 to 1.5 m. Thus, for much of the generating period, the turbines will be operated at the maximum possible power output for the conditions of head and submergence applying at the time. This is contrary to normal practice at most hydroelectric schemes, where the resource is limited for most of the time and so there is great emphasis on running the turbines at their maximum efficiency.

At D, when the difference in water levels across the barrage equals the minimum turbine operating head, the turbines are shut down and flow ceases until the sea level has risen to that in the basin. The sluice gates and the turbine flow control gates are then opened to refill the basin and start the next cycle.

With simple ebb generation, there is relatively little scope for operating the turbines or sluice gates to modify the shape of the tide curve behind the barrage. The gates could be closed before high water if the height of the tide would cause damage, as could be the case if there were a storm surge which resulted in an abnormally high tide. In the same way, flooding upstream by exceptionally large river flows could be reduced or prevented by shutting the gates early, thus limiting the level of high water in the basin. Towards the end of the generating period, the sluice gates, if so designed, could be opened early in order to drain the basin down to a level nearer the normal low tide level. This could be of use for the occasional maintenance of port facilities or sea defence works. Done on a regular basis, the areas of intertidal mudflats could be kept nearer that which existed before the barrage was built; among other effects, this could benefit wading birds and has been suggested for the Mersey barrage (Ref. 1988(7)). In addition, in a long estuary where dynamic effects are important in producing large tidal ranges, there may be a small benefit in energy output by reducing the water level in the basin before the next flood tide, because the incoming tide will encounter a smaller volume of still water and so could sweep into the basin to reach water levels closer to those occurring naturally. However, the resistance presented by the barrage will prevent the full natural levels being reached.

Because ebb generation is, with flood generation, the simplest mode of operating a tidal barrage, the choice of method of regulation of the turbines is widest. This aspect is discussed in the next Chapter.

2.3 Flood generation

This somewhat sinister term describes the mode of operation where the turbine discharge during power generation is in the same direction of flow as the flood tide, i.e. from the sea towards the enclosed basin. Fig. 2.6 shows a typical operating cycle for both ebb and flood generation for a $M2$ tide for a two basin Severn barrage as predicted by a 2-D model (Ref. 1981(17)). This shows that flood generation is virtually a mirror image of ebb generation and of Fig. 2.5,

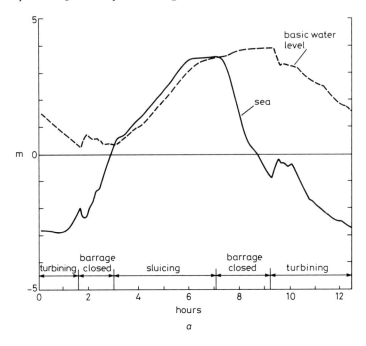

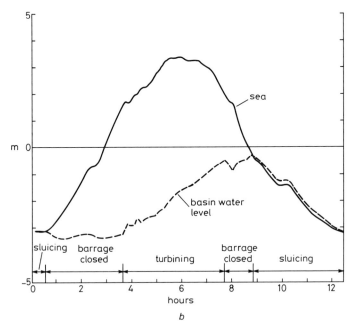

Fig. 2.6
a Ebb generation
b Flood generation

and therefore does not require detailed description. Otherwise, there are some important differences in the two cycles. Firstly, the range of water levels in the enclosed basin lies between normal low water and slightly above normal mid tide. This will leave a large proportion of the intertidal areas permanently exposed. This will result in a fairly rapid change from a marine ecosystem to a freshwater ecosystem, as rainwater dilutes and displaces the saline water in the foreshore deposits. The reduction in top water level will also have major repercussions for any shipping using the estuary.

Secondly, the volume of water in the basin between low water and mid tide will be much less than the volume between mid tide and high tide. The former is the useable volume of water for flood generation, the latter applies to ebb generation, and so flood generation is at a distinct disadvantage.

The only circumstances in which flood generation could be considered feasible would be in a steep-sided estuary where there would be a benefit in a permanent lowering of water levels; for example by improving land drainage. Otherwise, flood generation can form part of a two-way generation cycle. This is discussed later.

2.4 Ebb generation plus pumping at high tide

This is a variation of ebb generation, as shown in Fig. 2.7 to compare with Fig. 2.2. The basic concept is attractive. At or soon after high tide, when the sluice gates have been or are being closed, the turbines are operated in reverse so that

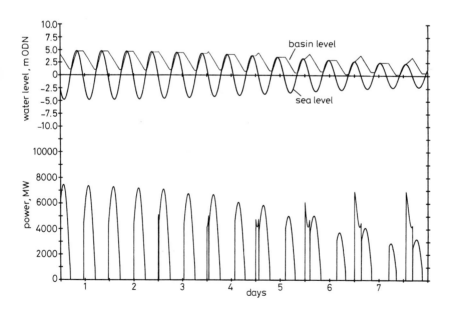

Fig. 2.7 Ebb generation with pumping at high tide between midnight and 8 a.m.

they act as pumps moving water from the sea into the basin. Pumping stops when the sea has ebbed enough to increase the head across the turbines to an uneconomical height. With more water stored in the basin, generation of power will normally start earlier than would be the case for simple ebb generation. The additional water in the basin is released through the turbines at a higher head than that through which it was pumped, so that there should be a net gain in energy. More important will be the cost of the electricity used for the pumping, which has to be imported to the barrage, compared with the value of the additional electricity generated. In the United Kingdom, if the pumping can be carried out just before the morning rise in demand, there could be a beneficial change in the value of the electricity between the end of the pumping and the start of generation.

An important factor here is that, for large tidal ranges, the turbines would have to be operated at their maximum output for most of the generating period to extract the maximum energy. Thus there is little scope for varying turbine operation to take advantage of a tariff change during the generation period, because any such variation will result in a reduction of total energy output. Consequently, the extra water produced by pumping can be used only at the beginning of the generation period by starting the turbines earlier. The scope for taking advantage of a tariff change thus becomes limited; the tariff change should take place between one and two hours after high water, i.e. perhaps during tides which have high water at around 6 a.m.

With smaller tidal ranges, there is more scope for varying the operation of the turbines and thus more scope for taking advantage of a tariff change.

At any point on the coast, there will be a close relationship between the tidal range and the time of high water. Table 1.2 in the previous chapter shows this relationship in the Severn estuary, from which it can be seen that high water of large spring tides occurs around 8 a.m. and 8 p.m. Similarly, high water of neap tides occurs around 2 a.m. and 2 p.m. In the UK, advantageous tariff changes occur at around 8 a.m. After allowing for the time that has to elapse between high water and the start of generation, there is a limited number of mid-range tides where advantage can sensibly be taken of these changes by pumping at high tide.

This conclusion applies to the Severn estuary. Elsewhere, high water of spring tides will be concentrated around a different time of the day, so that tides which will allow advantage to be taken of tariff changes will lie in a different part of the spring-neap cycle.

There are further factors to be considered with pumping at high tide. Firstly, since it has been demonstrated that the number of tides where a beneficial change in tariff can be exploited is limited, the turbines should be designed so that there is little or no reduction in the efficiency of the turbines when operating in normal ebb-generation mode; otherwise there will be a loss in overall output. The efficiency of the turbines is related to the method of regulating the water flow, discussed below, and the shape in section of the runner blades. For maximum efficiency in one direction of flow, the blades are curved in section rather like an aerofoil. For pumping in the reverse direction, either relatively low efficiency has to be accepted, or the blade shape has to be made more symmetrical, in which case efficiency during generation is lower.

The operation shown in Fig. 2.7 was based on the use of turbines with no concession to pumping efficiency, so that pumping would be at low efficiencies. This is seen in the large inputs of electricity required for pumping at neap tides.

Secondly, for a given barrage design comprising a certain number of turbines and a selected generator capacity, there will be an optimum speed of rotation for the turbines which results in the maximum energy output. This speed will be related to the rated head of the turbines, i.e. the head at which the generator reaches full output. Because pumping takes place at much lower heads, but at the same speed as generation, the efficiency of pumping will be well below the maximum for that design of turbine unless the turbines are equipped with variable speed control during pumping.

Thirdly, although the turbines for simple ebb generation could have either their runner blade angle or their distributor vane angle adjustable and still achieve high efficiencies, for pumping the runner blades must be adjustable to achieve reasonable efficiency and so the turbines must have either variable distributors as well (double-regulated), or downstream flow-control gates, in each case to control starting and stopping. Thus turbines suitable for reverse pumping are more complex and more expensive, although in the case of double-regulated machines they will also be slightly more efficient when generating.

Lastly, there may be benefits not related to energy production from pumping at high tide. The most obvious one will be the increase in water levels at dock entrances within the basin for large ships, ships which could otherwise be prevented from using that port by the reduction in natural high water levels caused by the barrage. Less obviously, the shorter period at high tide when water levels are constant could benefit wading birds which feed along the water's edge. The importance of these benefits is, of course, related to the proportion of the number of tides when pumping at high tide which will show an economic benefit or, at least, no disbenefit.

As with simple ebb generation, the dynamic effects of pumping from the sea into the basin can be important. Studies carried out using a 1-D model and reported in Ref. 1987(14) showed that the apparent gain in net energy output estimated with a flat-estuary model, which could be up to 10%, decreased to about 2% when dynamic effects were included. There appear to be three main reasons. Firstly, soon after high water at the barrage, the tide ebbing from the upper estuary and flowing down towards the barrage causes water levels to rise at the barrage more quickly and thus increase the head against which the turbines have to pump. Secondly, this early ebbing prevents the discharge from the turbines flowing into the upper estuary, so that water levels are raised near the barrage rather than uniformly over the area of the basin. Thirdly, the water pumped into the basin causes the water level on the seaward side of the barrage to drop faster than normal, so that the head across the turbines again is increased faster than expected.

This result for the Severn estuary appears to conflict with the results for La Rance (discussed in more detail below), where a 10% gain in net energy output is achieved by reverse pumping. There appear to be two reasons for this difference. Firstly, the output at La Rance in simple ebb-generation mode could have been larger, by optimising runner blade design, the design of the turbine water passage and the size of the generators. Secondly, the smaller gain in the

Severn may well be due to the much greater length of the basin behind the Severn barrage and thus a more dynamic regime than at La Rance which is closer to a 'flat estuary'.

2.5 Two-way generation

At first glance, the modes of operation shown in Figs. 2.2 and 2.6 could be combined to produce twice the energy output without increasing the cost of the barrage. In addition, the tidal range behind the barrage would be much closer to the natural range and therefore should be more acceptable to environmental interests. In practice, the operating cycle for two-way generation will look like Fig. 2.8. Starting at point A, the main features of the cycle are as follows.

Ebb generation starts at a basin water level which is much lower than that for single-cycle working, for reasons which will be explained later. At B, towards the end of the generating cycle, the sluices are opened in order to draw down the water level in the basin by allowing flow from the basin to the sea. This has to be done if a reasonably large difference in water level is to result during the next, flood generation, phase. After low tide, at C, the time when sea and basin levels are equal, the gates are closed and the basin level remains constant until the optimum difference in level is reached for the next phase of generation.

The turbines are now operated in the reverse direction of flow, flood generation. Again the difference in level between the sea and the basin is less than would be the case for one-way generation. At D, part way through the cycle, the sluices are opened again in order to raise the level of the water in the basin to a height which will result in a reasonable head across the turbines and therefore a reasonable amount of electricity being generated. It is the combination of lower heads during each phase compared with one-way operation, the

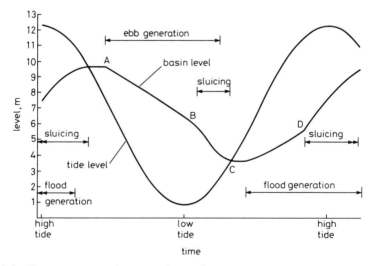

Fig. 2.8 Two-way generation operating cycle

'waste' of water by opening the sluices part way through each cycle, the smaller basin area, and thus useable volume, for flood generation, and the inefficiency of the turbines when operating in reverse flow that results in two-way generation not producing any more energy than one-way, ebb, generation. Although two-way operation has not been optimised with the aid of computer models in the same way that one-way generation has, studies have shown that, to produce the same energy, more turbines are needed and, as a result, two-way generation is about 10% more expensive per unit of electricity than one-way generation (Ref. 1981(1)).

Although two-way generation is less economic than one-way, there are certain advantages. The principal advantages are that the maximum power output is lower, because the maximum head across the turbines is lower, while there are two periods of generation per tide instead of one. Thus the output of a two-way generation barrage is much easier to absorb in the grid system. This becomes more important where the site is remote from the grid, and explains why several small schemes which have been built on the south-east coast of China, where there is not strong grid system, are two-way.

One possibly severe disadvantage of two-way generation lies in the large reduction in the level of high water in the basin. If the estuary is used by ships travelling to ports where water depths are critical, then the new levels will not be acceptable.

The 240 MW barrage at La Rance, near St. Malo in France, was designed to operate as a two-way scheme and also to be able to pump in either direction. The building and operation of this scheme have been well documented, e.g. Refs. 1978(5), 1984(3) and 1986(7). The decision to build it was taken at a time of uncertainty in France about the future development of the electricity system, and La Rance was intended as a large scale prototype test leading to the development of the very much larger Iles de Chaussee scheme (Refs. 1984(4)), 1986(15)). Thus the ability to operate in any of the four modes was seen as providing useful experience.

In the early years, La Rance was operated as a two-way scheme with pumping at high tide from the sea to the basin. The fourth mode, pumping at low tide from the basin to the sea, was used for less than 1% of the time. Fig. 2.9a shows this option (Ref. 1986(6)). In 1975 the stresses caused by starting the 10 MW generators as motors direct on line for pumping were found to have caused cracking of the stator fixings. Over the next seven years a programme of dismantling and repairing the generators in sequence was carried out, while the remaining units were not used for pumping. Subsequently, the barrage has been operated more as an ebb generation barrage with pumping at high tide (Fig. 2.9b). The advantages of pumping at high tide at La Rance appear to be partly due to the barrage having a relatively small area of sluices.

To conclude, the simplest and probably most economic method of operation of a tidal barrage is ebb generation. This allows the simplest turbines to be used and therefore should lead to the least maintenance costs. Ebb-generation with pumping at high tide is likely to be more acceptable to shipping interests and could have advantages for wading birds, but the number of tides where a clear gain in net energy or energy value could be obtained will be limited. Two-way generation is more costly per unit of electricity but the four periods of power

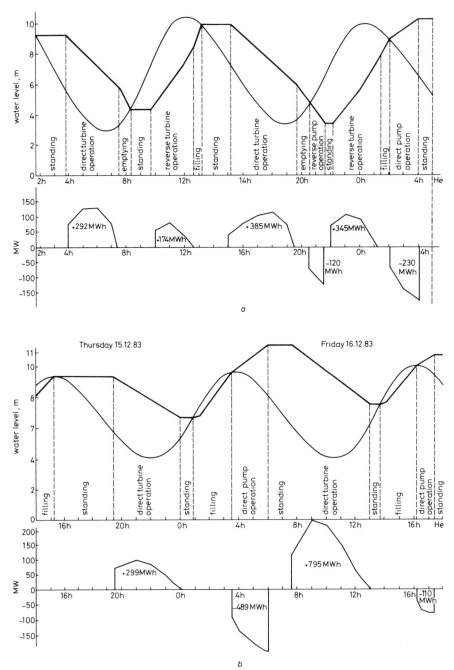

Fig. 2.9 Operating cycles at La Rance tidal power station
a Double effect operation
b Single effect operation

generation per day are easier for the grid to absorb or, where there is no grid, results in power being available for about twice the number of hours per day, which will be much more acceptable to the local community.

2.6 Two-basin schemes

If two basins can be formed adjacent to each other and each is equipped with a set of sluices of appropriate area, then a wide range of possibilities is opened up. The simplest development would be to have turbines located in the dividing wall, as shown in Fig. 2.10a. Basin A becomes the high level basin, filled through sluices at high tide. Basin B is the low level basin, emptied through sluices at low tide. The storage available in each basin allows the turbines to operate longer than is the case for a single basin scheme. If the capacity of the turbines is kept small, then continuous operation is possible, even through neap tides, but the total output of the scheme will be less than that of a single basin with the same area as the two basins.

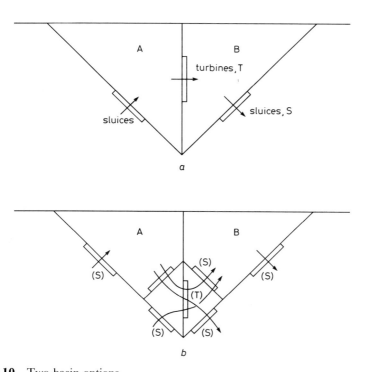

Fig. 2.10 Two-basin options
a High to low basin operation
b Multiple-path operation
   A = high level basin
   B = low level basin

A scheme of this type, and other versions, have been considered briefly for the north-west coast of Australia, where the topography happens to be suitable. This is described in Chapter 12. This type has also been considered by the Societé d'Etudes et d'Equipments d'Enterprises (SEEE) for EDF (Ref. 1987(6)). In general, the high cost of the civil engineering works offsets the benefit of more uniform output.

The next alternative would be to move the turbines towards one end of the dividing wall between the two basins, and install additional sluices between each basin and the turbines (Fig. 2.10*b*). This allows the turbines to be fed either from the sea into the low basin, or from the high basin to the sea, or from the high basin to the low basin. This can be beneficial in increasing the flexibility of the scheme to generate power at times when this is more valuable than normal. However, each time water passes through a set of sluices, or turns a corner, energy is lost in friction. This becomes more significant during neap tides when the available heads are small. The cost of the additional sluices and the additional head losses have to be offset against the increase in flexibility.

Another option is to design some or all of the turbines as pump turbines, or to equip one of the basins with pumps to enable it to be filled to a higher level than normal. This introduces an element of pumped storage. The latter has been

Table 2.1   Results obtained by SEEE for three schemes

	Single basin (Parametric)	Single basin (SEEE)	Double basin single path	Double basin multiple path
Area (km^2)	200	200	200	200
Assumed length of embankment (km)	20	—	—	—
Assumed maximum depth of water (m)	20	—	—	—
Assumed turbine diameter (m)	8	—	—	—
No. of turbines	72	54	52	40
Total generator capacity (MW)	2670	2160	2080	1600
Maximum yearly output (GWh)	5560*	5590	5580	5950
Capital cost (£M)	2900	1050†	1210†	1100†
Unit cost of energy (p/kWh)	4·0	—	—	—

* Assumed to be (average yearly output) × 1·06
† July 1980 prices in FFr updated to 1983 by 15% and converted at FFr10 = £1.00.

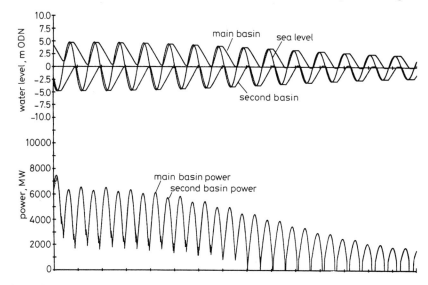

Fig. 2.11   Two basin scheme, one basin ebb generation, one flood generation

considered for the Australian scheme discussed in Chapter 12, and is included in the range of schemes studied by SEEE. In Ref. 1987(6) is a summary and comparison of energy outputs and costs for schemes with a total basin area of

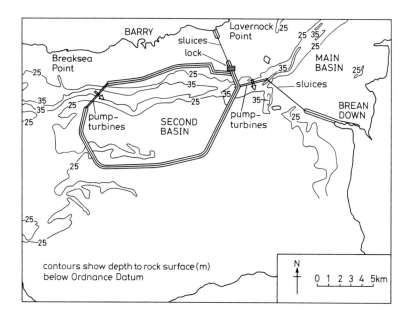

Fig. 2.12   Outline design of two-basin pumped storage scheme

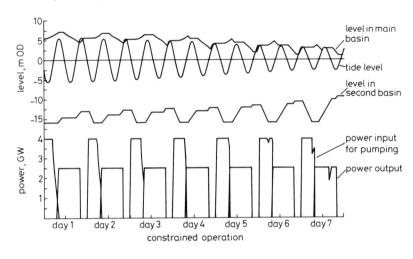

constrained operation

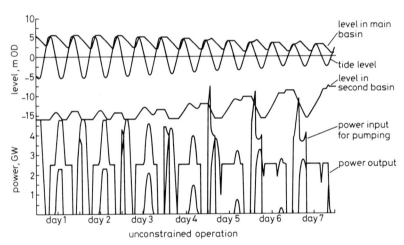

unconstrained operation

Fig. 2.13   Modes of operation of a two-basin scheme

200 km² location on the Cotentin coast between Jersey and the Chaussee islands, where the mean tide range is about 8 m, the same as at La Rance.

Using a parametric method for estimating the performance and cost of a single basin tidal power scheme (Chapter 11), the results obtained by SEEE can be compared with those for a single basin scheme of the same area, tidal range and so forth. These are set out for three schemes in Table 2.1.

This comparison shows that the energy output is close to that to be expected from a single basin, ebb generation scheme, even with a substantial reduction in installed capacity and with more complicated routes for the water flowing through the turbines. The estimated costs are, however, much less than the approximate estimate given by the parametric method.

The report of the Severn Barrage Committee (Ref. 1981(1)) discusses two types of two–basin scheme. The first comprised two separate basins, one a normal ebb–generation scheme, the other a normal flood–generation scheme. Fig. 2.11 shows how such a scheme would operate, with four periods of generation per day but still discontinuous output. Two sites were considered, the inner estuary, landward of the Holm Islands, and the estuary seaward of the preferred site of the Severn barrage. In each case, the long lengths of embankment required rendered the flood–generation basins uneconomic.

The second type of two–basin scheme comprised two basins each equipped with pump–turbines (Fig. 2.12). This had been proposed by Shaw (1980(14)) as a scheme capable of producing a substantial amount of power at a uniform rate during the day, while absorbing power at night to raise the water level in the upper basin by pumping from the sea into the basin at around high water, and lower the water level in the lower basin by pumping from the basin into the sea at around low water. Two methods of operation are shown on Fig. 2.13, referred to as constrained and unconstrained. The former aims to act as a pumped storage scheme with high efficiency, absorbing a block of power at night and generating a block of power during the day. This method of operation was aimed at an electricity supply system which included a high proportion of nuclear power plant which is difficult to cycle to meet varying demand. Detailed studies of this method of operation are reported in Ref. 1980(15). As can be seen, the water levels in the main basin remain at around or above normal high water level all the time, so that there would be substantial changes in the estuary as regards the exposure of intertidal feeding areas for wading birds, land drainage and flood evacuation. The second method, 'unconstrained', uses the main basin as a normal ebb–generation scheme with reverse pumping when appropriate, with the second basin tidying up the output of the main basin by pumping or generating at appropriate times. Studies showed that, although feasible technically, the pumped storage element, mainly the second basin, was about three times more expensive than a 'normal' inland high head pumped storage scheme and therefore uneconomic (Ref. 1981(1)).

Chapter 3
Turbines and generators

3.1 Introduction

The design and manufacture of the turbines and generators which are suitable for tidal power schemes is a commercially sensitive subject, with manufacturers understandably reluctant to disseminate their expertise and experience. This chapter therefore can deal with the broad principles only.

The operating range of turbines in a tidal power scheme will depend mainly on the spring tidal range at the site. The maximum head across the turbines normally occurs at low water, when the basin level has dropped some distance from high water, and will therefore be less than the tidal range. Thus the maximum head will typically lie in the range of 5 to 10 m. The upper end of this range also covers many run-of-river schemes and much relevant experience is available from this source. Having said that, the operating conditions for turbines in a tidal barrage are much worse than those in run-of-river schemes. The water is salty, the head is constantly varying, this variation possibly being exacerbated by storm waves on the seaward side of the barrage, and the operating cycle lasts a maximum of about 6 hours out of 12.4, so that the generators suffer thermal cycling. Suspended sediments could be present in the water, eroding metal parts and damaging shaft seals, although this can also occur on run-of-river schemes.

3.2 Development

The first turbines used for tidal power were the water wheels in tide mills. The builders had two choices regarding the basic design, the axle could be mounted either horizontally or vertically. If the axle was horizontal, then the bearings were above water, which made for easier maintenance, but the wheel had to be of large diameter to accommodate the variation in water level during the 'generating period'. If the axle was vertical, the diameter of the wheel could be relatively small but the lower bearing was under water and the control gear directing the flow of water onto the blades was relatively complex.

When serious consideration began to be given to tidal power, effectively starting with the appointment of the first Severn Barrage Committee in 1927 (Ref. 1933(1)), it was logical to rely on the experience that had been gained on low-head hydro projects, where the most developed turbine was the vertical axis Kaplan turbine. This design offered the advantages of good efficiency over a wide range of flows and heads and, with the generator mounted in the dry above

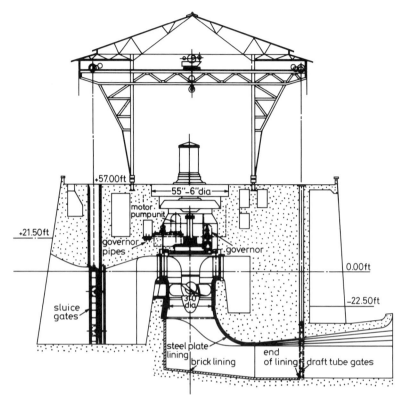

Fig. 3.1   Arrangement of turbine proposed for Severn barrage in 1945

the turbine, easy cooling and good access for maintenance. Fig. 3.1 shows a typical arrangement of this type of turbine as proposed for the Severn Barrage in 1945 (Ref. 1945(1)). The main disadvantage lies in the water passing through the turbine having to turn through 90° twice. This would waste some of the relatively low head available.

In order to reduce these losses, three designs were developed, namely the rim generator turbine, the tubular or 'S' turbine and the bulb turbine.

3.3 Rim generator turbine

The concept of the rim generator turbine was patented by Harza, an American engineer in 1919. The current patents are held by Sulzer Escher Wyss of Zurich, who have manufactured about 100 small machines of this type, latterly under the trade name Straflo.

The concept is illustrated in Figs. 3.2 and 3.3. The turbine runner is of Kaplan propeller type, mounted horizontally unless site conditions dictate otherwise. The generator rotor is supported on the tips of the runner blades and

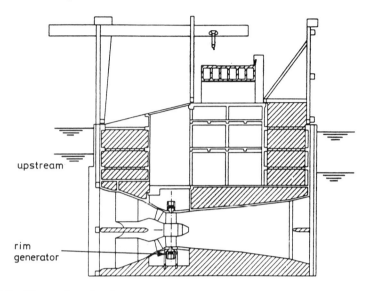

Fig. 3.2   Illustration of a rim type generator

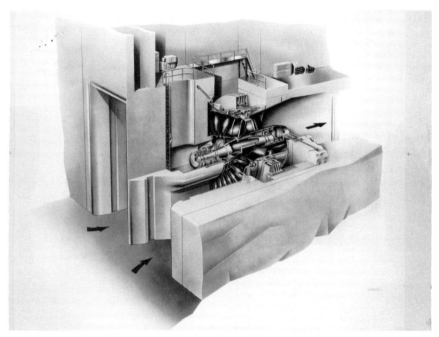

Fig. 3.3   Large rim generator or 'Straflo' turbine
(Sulzer-Escher Wyss, Zürich)

outside the rotor is the stator. This arrangement has the advantage that the design of the generator is not constrained by lack of space, and cooling is relatively straightforward. In addition, the mass of the rotor mounted on the outside of the runner provides high rotational inertia, which assists operational stability.

A disadvantage of this design lies with the sealing of the gaps between each side of the rotor and the water passage. This sealing not only has to accommodate flexing of the runner blades under load, but also temperature changes and a high relative speed between fixed and moving parts.

Until 1984 the nearly 100 machines which had been built had runner diameters up to 3.7 m. Then a prototype 20 MW machine with a runner diameter of 7.6 m was commissioned at Annapolis Royal, a small inlet off the east side of the Bay of Fundy in Nova Scotia (Refs. 1986(2), 1987(4)). This machine was designed by Escher Wyss, manufactured by Dominion-Bridge/Sulzer in Canada and transported to the site as three main assemblies on a barge. Fig. 3.4A shows the runner and rim-mounted generator rotor. Fig. 3.4B shows the distributor in transit. Fig. 3.4C shows the generator stator being installed.

Apart from representing the first development of tidal power in North America, this machine is also providing experience relevant to large run-of-river projects. After some initial problems with the insulation of the generator poles had been corrected, very high availability has been achieved. Two seals are provided around the sides of the rotor to exclude sea water from the generator. One is inflatable and is for use when the machine is stationary. This was used for several months when progress was stopped during construction by a strike, so that it stuck to the rotor and was damaged during the subsequent start-up. The second seal is a 'hydrostatic' seal which relies on maintaining a small gap with filtered sea water introduced under pressure, and the excess water is collected in suitable shrouding. This sealing arrangement, about which there was much interest before the project was built, has generally worked well, the main modification being an increase in the capacity of the filtration equipment to cope with suspended sediments in the estuary during the spring run-off and with marine organisms during the summer.

Fig. 3.3 shows a cutaway view of the Annapolis Royal machine. A different outline design was developed for the Severn Barrage studies (Fig. 3.2.). This does not have a bearing downstream of the runner; the runner bearings are on an axle which is cantilevered from the upstream support. This arrangement would make both the assembly and the removal for maintenance of the turbine and generator simpler and quicker. A second advantage could be that fish passing through the turbine would not be at risk from damage by hitting the downstream supports. This appears to be a feature of the Annapolis machine when it is running off maximum efficiency and therefore causing swirling flow downstream (Ref. 1988(6)).

Compared with the bulb turbine, discussed later, the rim generator turbine offers some economy in the overall length of the water passage because it is shorter. However, this advantage may be offset by the overall diameter, and hence the spacing of adjacent turbines, being larger than for bulb or tubular turbines.

Fig. 3.4
a Runner and generator rotor for Annapolis 'Straflo' turbine
(Sulzer-Escher Wyss, Zürich)
b Distributor for Annapolis 'Straflo' turbine
(Sulzer-Escher Wyss, Zürich)

Fig. 3.4
c Installation of stator at Annapolis Royal
(Sulzer–Escher Wyss, Zürich)

3.4 Tubular turbines

Fig. 3.5 shows a typical outline design of a tubular turbine, as prepared for the Severn Barrage (Ref. 1981(1)). The purpose of the design is to move the generator from the water passage into a dry enclosure. This is achieved by means of a long shaft connecting the Kaplan runner at an angle through the wall of the water passage. Because there is plenty of space around the generator, the speed of the generator can be increased by inserting a gearbox between it and the runner. This can result in significant reduction in cost, because a generator designed to run at a speed of say 600 rpm will be near standard, while a generator designed to run at 60 rpm, which is a typical speed for a large tidal power turbine, will be large and expensive.

Tubular turbines are offered by several manufacturers in sizes up to about 10 MW. The largest machines of this type built so far are the five 8 m diameter

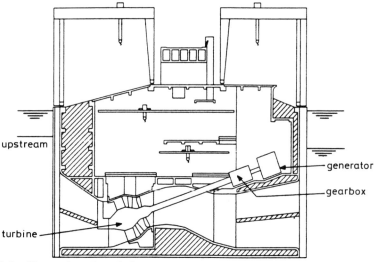

Fig. 3.5   Illustration of a tubular type generator

25 MW machines for the Ozark project in the United States, commissioned in 1970 (Ref. 1978(4)). These included a speed-increasing gearbox using spur gears. Problems were reported with the flanged connections at the gearbox end of the long drive shafts and with vibrations at high power outputs. Apparently, no large tubular turbines have been built since the six 6.45 m diameter machines for the Harry S Truman Dam in the United States which were commissioned in 1976.

At the present time, no serious consideration is being given to the use of tubular turbines for tidal power because of the general superiority of the bulb type.

3.5   Bulb and pit turbines

Bulb turbines have their generators enclosed in watertight steel bulbs upstream of the runners. Pit turbines are similar but have their generators within 'pits' which are open-topped, with the water passage split to pass each side. The runners are again of Kaplan type. Fig. 3.6 shows a typical arrangement of a bulb turbine with directly-driven generator. A large number of bulb turbines have been manufactured over the last 50 years for low-head and run-of-river schemes. The largest runner diameter built so far is 8·4 m for two 20·5 MW machines, built by J M Voith GmbH for the Murray lock and dam near Arkansas, USA. These are pit turbines. The largest runner for true bulb turbines is 8.2 m for the Vidalia scheme, for which four out of the eight runners were manufactured by Markham in Chesterfield, UK (Fig. 3.7). The largest generator capacity used so far is 65 MW at 60 Hz, built by Hitachi of Japan (1986(4)). One of the largest installations comprises eight turbines with 7·4 m diameter runners and generators rated at 54 MW at 60 Hz and 85·7 rpm

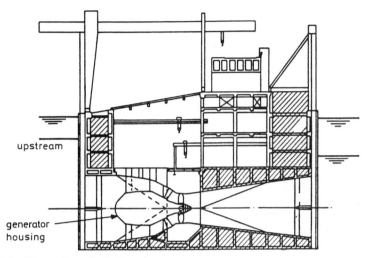

Fig. 3.6 Illustration of a bulb unit

supplied by Neyrpic of France for the Rock Island project in the United States (Fig. 3.8).

The Rance turbines are of bulb type, with a runner diameter of 5·4 m and with 10 MW generators at 93·75 rpm. The success of these machines, apart from a problem with the fixings of the stators within the bulbs discussed in the previous chapter, combined with the great amount of experience with all sizes of

Fig. 3.7 Runner for Vidalia scheme (Markham & Co)

Fig. 3.8   Rock Island Project

machines on non-tidal projects, means that this type is generally favoured for large tidal power schemes. There are other factors which are discussed below.

Recently, the advent of epicyclic gearboxes rated at several megawatts at the torques appropriate to large bulb turbines has led to several machines being built which include an epicyclic gearbox within the bulb to increase the speed of the generator. Perhaps the most important are the eight 25 MW machines for the Vidalia project. This is particularly advantageous, because the space within the bulb has to be severely limited if the water flow pattern around the bulb is not to result in undue hydraulic losses. This constrains the design of the generator and its cooling system.

A few bulb turbines have had their generators water cooled. The majority, including the more recent large machines, have been air cooled. It is more or less standard practice for the air inside the bulb to be pressurised to about two atmospheres. This increases the density of the air and so improves its efficiency as a cooling medium, and at the same time reduces the pressure difference between the inside of the bulb and the surrounding water. The air is forced through passages in the rotor and stator, and then cooled by passing through water cooled radiators which transfer the heat to the outside of the bulb for dissipation into the surrounding water.

3.5 Turbine regulation

If the flow of water and head across a turbine were constant, then the power available to be extracted from the water would be constant and the turbine would need no mechanical control of water flows; the speed of the turbine would be held constant by controlling the generators to produce the correct power. Thus the runner blades would be set at a fixed, optimum, angle.

Upstream of the runner is a 'distributor', which is a ring of blades which imparts a spiral motion to the water before it reaches the runner blades so that the angle of flow relative to the blades is also an optimum. In the constant operating conditions postulated, the distributor blade angle can be fixed as well. Such a turbine is 'unregulated' in terms of water flow, and since both the runner blades and the distributor guide vanes are fixed, these are relatively simple to make and capital and maintenance costs are reduced.

If control of the turbine is lost for any reason, for example load–shedding, then the turbine will 'run away', its speed and the flow of water both increasing rapidly until limiting conditions of water velocity are reached. Since, under these conditions, the hydraulic efficiency of the turbine will be minimal, operation under runaway is rough and noisy, and is avoided carefully. The turbine then has to be slowed down by closing a downstream gate, itself designed for the resulting large loads. This gate is also used to control water flow during start-up until the turbine has reached synchronous speed. It has to be located downstream to avoid asymmetrical flow at the turbine.

Unregulated bulb turbines have been built, perhaps the most noteworthy examples being four machines at Caderousse in France. These have 6·9 m diameter runners and generators rated at 31·5 MW at 93·8 rpm. They act as base load machines, variations in river flow being taken up by two double regulated machines of the same rating but with 6·25 m runners.

Double regulated machines have both their runner blades and their distributor vanes adjustable. This enables them to maintain high efficiency over a wide range of water flow and head. In addition, the distributor vanes, up to 20 in number, are normally designed to be able to close completely. This is useful for a tidal power scheme where the turbines have to be started and stopped every tide. The normal number of runner blades is four, but at least one recent large bulb turbine built in Japan has five blades. It may in due course be feasible to design the runner blades and, perhaps, increase their number to six, so that they could be closed completely and thus be used for starting and stopping the turbines.

Turbines can be single regulated, with either adjustable distributor vanes or adjustable runner blades. Neither is common, the great majority of bulb turbines being double regulated. For a tidal power scheme operated in ebb-generation mode only, either type is feasible, as is the double regulated type. However, for two-way generation double regulation is necessary. With one-way generation, bearing in mind the need to close the turbines at high tide, a single regulated turbine which has an adjustable distributor has the advantage of not requiring a downstream gate for starting and stopping, this being an expensive item which also complicates the design of the power house. On the other hand, although the maximum efficiency of the two types is similar, the operating band of turbines with fixed runner blades is narrower than that of machines with fixed distributors, and a fixed distributor is much simpler and cheaper to manufacture than a variable distributor. Figs. 3.4B and 3.8 show the outside of an adjustable distributor for a large bulb turbine and indicates the complexity of the operating mechanism.

If a turbine is to be designed to pump from the sea into the basin at around the time of high tide, then either a double-regulated design or a single-regulated machine with fixed distributor and adjustable runner blades (and downstream gate) is necessary. The former will be more efficient but also more expensive.

The concept of the Straflo turbine, with the generator rotor mounted on the tips of the runner blades, makes it difficult to make the runner blades adjustable. The Annapolis Royal machine has an adjustable distributor and, significantly, no downstream gate. In this the designers have taken advantage of the tidal cycle which ensures that a turbine must come to a stop twice each tide when water levels in the basin and the sea are equal. Thus, in the event of the distributor failing to operate properly, control of the turbine cannot be lost for more than about six hours, after which levels must equalise for a short time and a gantry can be used to insert simple stop–logs downstream to close off the flow. The same concept of a gantry with gate, ready to service a number of turbines, was adopted at La Rance and was proposed for the Severn barrage in Ref. 1981(1). This cannot be the case on a run-of-river scheme, where loss of control could result in a turbine running at overspeed indefinitely, and so a permanent downstream gate has to be provided for each turbine.

From these considerations, it follows that the choice of method of turbine regulation for a tidal power scheme is quite finely balanced unless two-way operation is planned, in which case double regulation is required, or pumping at high tide, which requires either double regulation or a machine with fixed distributor and variable runner blade angle (and downstream gate). Double regulated machines are most efficient and thus produce most energy, but are the most expensive. Single regulated turbines with adjustable distributor vanes avoid the need for downstream gates but are slightly more sensitive to operate, and the distributor mechanism is expensive and complicated. Single regulation by adjustable runner blade angle is an option not yet available with large Straflo turbines. It requires a downstream gate to start and stop the turbine, but the fixed distributor, which needs only six large vanes, is simple, robust and contributes to the support of the turbine.

3.7 Design, manufacture and installation

Rarely does a single low-head hydro-electric project have more than eight or ten turbines. La Rance has 24 turbines and thus is exceptional, but the large quantities of water which are available each tide mean that large tidal power schemes require large numbers of turbines. The Severn barrage could require about 200 of the largest turbines yet built. Bernstein (Ref. 1986(3)) has proposed schemes in the USSR which would require up to 1500 turbines. The turbines for the Severn barrage would be required to be manufactured over a period of about four years, i.e. one turbine and generator every week. An order for this number of turbines and generators would not only set the appropriate parts of the UK manufacturing industry on its feet, it would also 'rock it back on its heels'.

The materials for bulb turbines include stainless steels for the runner blades, the distributor vanes and perhaps the lining of the water passage adjacent to the runner. The main shafts are normally carbon steel and the bulb skin is structural steel plate. These and other exposed steel parts have to be protected by an impressed-current cathodic protection system. This type of system has been very successful at La Rance and early indications are that it is working well at Annapolis Royal.

So far, large bulb and rim generator turbines have been assembled in position. This is a slow process and appropriate features have to be incorporated in the design. An example is the distributor, the main ring of which could be over 10 m in diameter and therefore too large to be transported by road or rail. Therefore, these are designed with radial joints which complicate design and manufacture and add to the cost. If a large tidal power scheme proceeds beyond the study stage, there will be scope for designing, manufacturing and installing the turbines and generators (and their ancillary equipment) around the concept of a purpose-built production facility adjacent to a suitable dockside. These facilities could take advantage of a large production run by having numerically controlled machining instead of hand finishing, by having jigs to ensure accuracy and interchangeability (and perhaps dummy turbines at the caisson construction facilities to ensure that the concrete structures were built accurately where necessary) and by removing the constraints on manufacture that arise when large elements have to be transported overland.

Turbines and generators have been assembled in their caissons away from the barrage site and the completed structures floated into position. The first example was the small 400 kW experimental tidal power plant which was installed at Kislaya Guba, an inlet off the White Sea on the north coast of Russia, in the early 1960s (Ref. 1965(2)). The turbine has a runner diameter of 3.8 m and drives the generator through a step-up gearbox.

More recently, Alsthom Atlantique and Neyrpic built three 26 MW bulb turbines and installed them in two steel caissons which were then carried across the Atlantic on a submersible load-carrying ship, off-loaded and towed up the Mississippi and Ohio rivers to be built into an existing dam to form the W.T. Love project (Ref. 1983(2)). When the caissons were sunk in their final positions within a temporary cofferdam, the cofferdam was dewatered and the caissons concreted in. Thus this project was not fully prefabricated in the way proposed for the Severn barrage. Nevertheless, it demonstrated the application of the general method where the construction period was short and periods of severe river flooding had to be avoided. In 1987, Boving & Co were awarded a contract for a single steel caisson containing eight turbines with 8.2 m diameter runners, again to a tight programme, this being the Vidalia project, already mentioned and also in the United States.

Studies for the second Severn barrage Committee (Ref. 1981(1)) showed that the turbines for the Severn barrage should not be installed in the caissons before float-out. The main reason is that the caissons should be installed as fast as they are completed so that the barrage can be put to work as soon as possible, using as many turbines as are available at the time. For the best return on investment, about half the turbines should be ready for operation at the time that the barrage structure, including sluice gates and so forth, becomes operational. This highlights the need for turbines to be designed and built in a way that enables the amount of work on the barrage to be kept to a minimum. Out of this

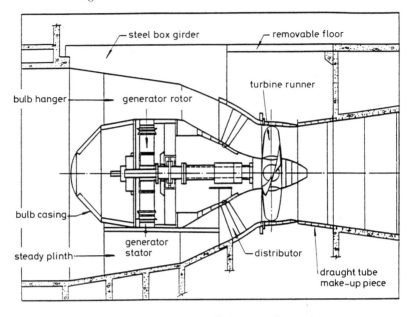

Fig. 3.9   Bulb turbine designed to be installed as complete unit

developed the concept of turbines and their generators being built as complete units at the special shore-based facilities discussed above, loaded onto a special barge and lifted into position at the barrage by a heavy-lift crane.

Fig. 3.9 illustrates this concept which is very similar to that of a jet engine within a pod slung below an aircraft wing. The complete bulb turbine with its distributor are supported by the access tower from a box girder structure which forms the roof of the water passage over the turbine. Additional location fore and aft, and sideways, is provided by pins in a streamlined support block underneath the bulb, similar in shape to the access tower. A crucial factor will be the resistance of the complete installation to vibration caused by moving parts being dynamically out of balance, particularly the runner which is furthest from the support. Preliminary finite element analyses (Ref. 1981(15)) have shown the concept to be feasible.

The weight of an 8 m diameter bulb turbine built to be lifted into position as a complete unit would be over 1000 t. This is well within the capacity of a number of floating cranes built for the offshore oil industry and so would not require the development of new technology. However, the rapid rise and fall of the tides would complicate normal lifting and precise lowering of such heavy items, and because there would be such a large number of identical lifts, it may be economic to build a purpose-designed crane, able to carry the turbines if necessary and perhaps with a jack-up capability. Ref. 1986(5)) highlights the care and time needed to load and offload an item of plant weighing about 1000 t without the benefit of a heavy lift vessel.

3.8 Turbine performance

As stated at the beginning of this chapter, the performance of turbines is a commercially sensitive subject. The designer of a tidal power scheme has to approach manufacturers direct for assessments of the most appropriate design for the particular circumstances of tidal range, water depth and available volumes of water that apply to that site. The extent of the advice received will depend on the manufacturer's view of the prospects of the scheme proceeding and the extent of the competition for the supply of the turbines and generators. This is only to be expected—turbine design and manufacture is a highly competitive international business.

Turbine performance figures are based largely on the results of model tests, using models with runner diameters of around 300 mm. The advent of computers, three-dimensional flow models and finite-element analytical methods means that much work can be done by computer that used to be done by relying on the judgment and experience of the turbine design engineer and previous model test results. However, scale models still have an important role to play, especially when the turbine involves novel aspects of design or performance, because the test results will form the basis of the guaranteed performance of the turbines, covering efficiency, cavitation behaviour, runaway speed, guide vane operating torques and so forth.

The model tests will also include the proposed geometry of the inlet to the water passage and the draft tube downstream. These have an important bearing on the overall performance of the turbine, since they convert the potential energy of the difference in water levels on either side of the barrage to kinetic energy at the turbine, and then convert most of the residual kinetic energy of the water leaving the runner back to potential energy. This reconversion reduces the head losses downstream of the turbine and so makes a greater proportion of the total energy available for extraction by the turbine. It follows that the length of the water passage is one aspect of the design of the complete barrage which has to be balanced against cost of construction and energy benefits, with the proviso that shortening this length incurs a risk of instability in the flow of water. Ref. 1981(12) discusses this aspect of turbine design in connection with normal hydro-electric schemes. Fig. 3.10 shows the proportions adopted for the water passage for the Severn barrage in Ref. 1981(1). These were based on published data for run-of-river schemes modified to allow for the concept of a concrete caisson. The sharp edge to the inlet would have to be rounded to improve flow.

The performance of a model turbine is normally presented in dimensionless form, essentially the performance of a turbine with a runner diameter of 1 m. Because a wide range of operation is possible, the performance is defined as a 'hill chart' relating discharge, head and efficiency. Fig. 3.11 is an example, defining the performance of a particular design of double regulated turbine prepared by Sulzer Escher Wyss of Zurich and widely used in recent tidal power studies. For reasons of confidentiality, the actual percentage efficiencies have been replaced by the percentage of maximum efficiency.

The vertical axis of Fig. 3.11 defines the specific discharge of the turbines, Q_{11},

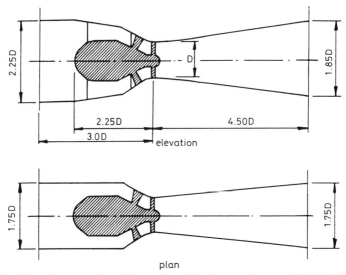

Fig. 3.10   Draught tube dimensions adopted for Severn barrage (Ref 81(1))

which relates the flow and head in the form:

$$Q_{11} = \frac{Q}{D^2 H^{0.5}}$$

The horizontal axis defines the head in terms of the unit speed n_{11} in the expression:

$$n_{11} = \frac{nD}{H^{0.5}}$$

The curves of various angles represent the degree of opening of the runner blades and the distributor vanes, and are not of immediate interest. For a tidal power scheme, studies have shown that the turbines should be operated for most of the time at the maximum output possible at the head available. In Fig. 3.11, this is defined by the curve labelled 'max output', ABC. The part AB represents the maximum allowable opening of the turbine distributor and runner blades for acceptable operation, and is therefore the maximum flow. The part BC represents the maximum product of flow, head and efficiency; above this line the increase in flow is more than offset by the reduction in efficiency. The line of maximum efficiency is not shown but is defined by the intersections of vertical tangents with the efficiency contours.

To demonstrate the use of this hill chart, an example is taken, namely a turbine with a runner diameter of 8 m and a maximum power output to the generator of 13 MW, suitable for a site where the head across the turbine can reach 5·5 m but the normal operating head for much of the time is likely to be around 3·5 m. For the purposes of this exercise the best efficiency of the turbine is taken as 80%; the actual, confidential, figure will be higher than this.

The starting point is to select a speed of rotation which results in the power output being about correct at the design point, namely the point where Q_{11} is

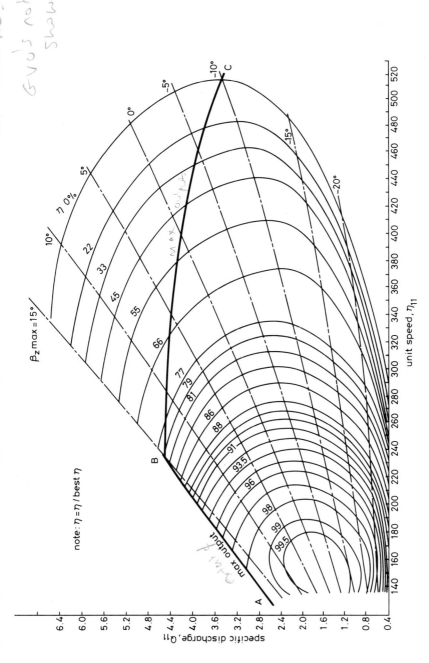

Fig. 3.11 Performance 'hill' chart for double regulated turbine

maximum. From Fig. 3.11, this point is where:

$$n_{11} = 235$$
$$Q_{11} = 4.5$$
$$\eta = 80\% \times 80\% = 64\%$$

To start with, the speed of the turbine has to be established by trial and error; too high a speed will give too much power (and also require more submergence, discussed below). A synchronous speed has to be selected, given by

$$\text{speed } n = \frac{2 \times 60 \times f}{N}$$

where f = frequency of supply system (Hz)
  N = number of generator poles.
Thus for a 60 Hz supply and 128 poles (an even number, 64 pairs of poles), the speed is 56.25 rpm.

  Substituting back in the equations for n_{11} and Q_{11} gives

$$H = \left[\frac{nD}{n_{11}}\right]^2 = 3.667 \text{ m}$$

$$Q = Q_{11}D^2H^{0.5} = 551.5 \text{ m}^3/\text{s}$$

$$\text{Power } P = \rho g Q H \eta = 1.03 \times 9.81 \times 551.5 \times 3.667 \times 0.64$$

$$= 13.078 \text{ MW}$$

where ρ = density of sea water (t/m^3)
  g = acceleration due to gravity (m/s^2)
This appears to be close to the required performance. The next stage is to check that there is adequate submergence of the runner to prevent cavitation. Cavitation is caused by the reduction in pressure as the water accelerates towards the runner, reaching a level where cavities form in the water, at a pressure equal to the vapour pressure of water at that temperature. These cavities usually appear first around the roots of the runner blades. When they collapse, owing to a slight pressure rise, they implode and result in high instantaneous local pressures which can be very damaging, even to high quality steel. Fig. 3.12 shows the relationship developed by Escher Wyss between cavitation limits, specific discharge, submergence and head, given by

$$\sigma = \frac{B^* - Hs}{H}$$

where B^* = atmospheric pressure less the vapour pressure of water (about 10.3 m at sea level).
  H_s = depth of submergence to root of runner blade
  H = head across turbine
  In the example chosen, for $Q_{11} = 4.5$, $\sigma = 5.4$ giving $H_s = -9.5$ m (i.e. 9.5 m below sea level).
For an 8 m diameter runner, the blades will be mounted on a hub with a diameter of about 3 m, so the axis of the turbine will have to be submerged to a

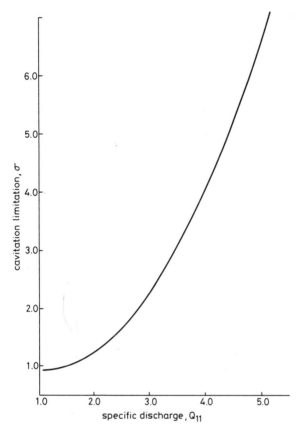

Fig. 3.12 Cavitation limitation

depth of about 11 m at low water of spring tides, which is when the maximum head is developed.

If this submergence is not available, the turbine parameters will have to be changed. Since submergence is related to Q_{11}, the answer is to reduce Q_{11} slightly at the design point and move down the maximum output line. For the example taken, the maximum value of Q_{11} which will not result in cavitation is obtained from

$$-H_s = \frac{\sigma H}{B^*}$$

whence $Q_{11} = 4 \cdot 3$.

Returning to Fig. 3.11, with $Q_{11} = 4 \cdot 3$, $n_{11} = 223$ and $\eta = 83\% \times 80\% = 66\%$, the synchronous speed which results in the maximum power being closest to 13 MW is then $53 \cdot 73$ rpm (134 poles at 60 Hz). The turbine parameters are then:

$$H = 3 \cdot 71 \text{ m}$$
$$Q = 530 \cdot 5 \text{ m}^3\text{s}$$
$$P = 13 \cdot 125 \text{ MW}$$

Table 3.1   Example of turbine performance

H	n_{11}	Q_{11}	$Q(m^3/s)$	$\eta(\%)$	$P(MW)$
0·74	500	3·6	198	10	0·15
1	430	4·1	262	32	0·85
2	304	4·36	395	58	4·63
3	248	4·36	483	64·5	9·44
3·71	223	4·3	527	67	13·2
4	215	3·5	440	74	13·2
5	192	2·3	332	79	13·2
5·5	183	2·0	299	79·5	13·2

With the basic turbine performance parameters defined, the performance over the full range of heads can then be extracted from Fig. 3.11, by following the maximum power line as closely as possible up to the chosen generator capacity and then reducing Q, the flow through the turbine, as necessary. Table 3.1 summarises the results.

Figure 3.13 summarises the performance of this turbine. The curves can be incorporated into computer models in the way that suits the program best. This can be by means of cubic equations, established using a standard curve-fitting routine, or as a series of points with the program interpolating as necessary.

Detailed studies of the method of operating the turbines for a tidal power scheme have shown that, in the absence of local dynamic effects caused by reflections of the flood tide wave at the top of the estuary, the path ABC shown in Fig. 3.11 results in the maximum energy being generated. At the start of power generation, operation starts briefly near the maximum efficiency line, travels upwards towards the maximum power line along a line of high constant efficiency and finishes by running down the maximum power line until the minimum head is reached. This highlights the difference between a tidal power scheme and a run-of-river scheme in terms of turbine operation: the latter will have turbines designed to operate at or close to the point of maximum efficiency. It follows that the ideal turbine for tidal power will have a broader band of high efficiency, perhaps at the expense of the absolute maximum being reduced somewhat. By making the turbine more efficient at low heads, pumping performance, which takes place at relatively low head, should also be improved.

Ref. 1981(14) sets out a logical method of optimising these variables on the basis of the unit cost of energy. Chapter 10 covers some of the same ground.

3.9 Air turbines

Because water turbines for tidal power schemes have to contend with a range of problems not normally encountered on run-of-river schemes, such as a wide range of heads and rapid thermal cycling, the possibility of using air turbines has been studied in some depth. These would work on the principle that the

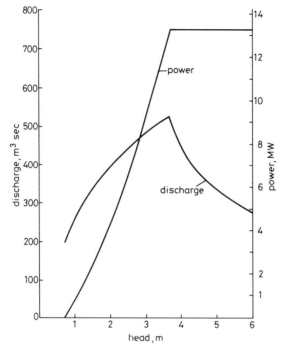

Fig. 3.13 Notional double-regulated turbine characteristics

water flow would be directed into some form of enclosed chamber where the water surface would rise and displace air. This air could then be run through an air turbine. By having two chambers side by side, the valving could be arranged so that the air displaced by a rising water level in one could transfer into the other, via a turbine, at the time that its water level was falling. The air turbine would be operating in velocities much higher than normal for a water turbine, thus allowing a much higher speed of revolution, such as 3000 rpm. This in turn makes the turbine and generator more compact and reduces their capital cost. The principal disadvantage lies in the density of air being only about 1/800 the density of water. Since the power available in a flow of liquid is the product of its flow rate, its density and gravity, converting from water to air before extracting energy must involve losses.

Y-ARD (1987(9)) have suggested a more complex arrangement of air turbines and compressors for the Severn barrage, with interconnecting ducts through which air flows would be controlled by valves. The overall efficiency of the arrangement proved disappointing and this proposal has not been taken further. An aspect of concern as regards the engineering of the project was the need for large gates opening and closing quickly and within a short cycle time to control the water flow.

Within the context of small scale tidal energy, a brief survey and comparison of alternative types of turbine is reported in Ref. 1987(9). This showed that water turbines of Kaplan type (bulb, pit and Straflo) are the most suitable for tidal power schemes with turbine diameters up to around 4 m.

Chapter 4
Caissons

4.1 Introduction

'Caisson' is the name given to a prefabricated structure which is floated into position. Since the early days of civil engineering, caissons have been widely used for forming bridge piers, a concrete or steel box being lowered onto the river bed and the base sealed into the ground, generally by skilled labourers working inside in compressed air. To construct the Ranee tidal barrage (Figs. 4.1 and 4.2), the area to be occupied by the barrage was enclosed by two temporary cofferdams and dewatered. The cofferdams comprised float-in cylindrical caissons of steel placed at close centres, with the gaps sealed with steel piling, and were therefore major structures in their own right. Completing them in the strong tidal flows presented severe problems. Then, because the estuary was largely blocked off for several years, the water in the enclosed basin first turned from sea water to almost fresh water, so that the marine life was unable to survive, and then, after the cofferdam was removed, returned to sea water. This disruption of the marine life in the estuary would not be considered acceptable now, so a tidal power scheme would have to be built in such a way that the estuary was not closed at any stage. One possibility would be to build the power house and sluices on dry land on the shores of the estuary and, when complete, open up inlet and outlet channels and then close the main estuary. This is the method that has been used for some small schemes in China (Chapter 12) and has been considered elsewhere (Ref. 1981(4)). This method requires much land. An obvious alternative is to use caissons for the main parts.

4.2 Experience with caissons

The caissons used for bridge piers were generally made of cast iron segments joined together by rivets or bolts. One of the earliest examples of a large prefabricated concrete structure floated into position is the Nab tower off the eastern end of the Isle of Wight. This was built during the First World War as the first of several towers designed to be placed at intervals across the English Channel to support anti-submarine nets. The war ended before the towers were all completed and so the Nab tower was floated into a position where it would be of use as a navigation landmark.

The classic early use of caissons which is relevant to tidal power are the Phoenix caissons which were towed across the English Channel to form the Mulberry harbour to support the allied invasion of Europe in 1944. This

Fig. 4.1 La Rance barrage, looking towards right bank

concept was a brilliant piece of lateral thinking around the problem of capturing a port in good enough condition to be of immediate use. 140 units were built in a period of about 18 months under wartime conditions, each unit weighing about 6000 t. After the end of the war, many were raised and re-used in the repair of the sea defences in the Netherlands, which had been breached to flood the land to prevent it being used to grow food for the invading Germans. A few can still be found and are generally in good condition, in spite of being designed for a short life and thus built with minimum concrete cover to the reinforcing steel. Fig. 4.3 shows two units which were re-floated into a new position in Portland harbour in the 1960s to form a wind break for a naval loading jetty.

Recently, as the final part of the defences against rising sea levels and rare, exceptionally high 'storm surge' tides, the Oosterschelde (or Eastern schelde)

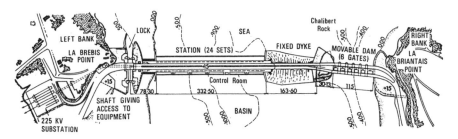

Fig. 4.2 La Rance Structures (plan)

Fig. 4.3   Two Phoenix caissons in Portland harbour, UK

storm surge barrier has been completed in the Netherlands. For this, 66 precast concrete piers weighing 10 000 t each were built in a special workyard formed by draining an area of the sea bed enclosed within an embankment of dredged sand. When the piers were complete, the workyard was flooded and the piers, partly buoyant, were lifted, transported to their required locations and placed accurately in position by a special catamaran/crane. The piers were then filled with concrete and the barrier completed with precast beams, which also formed a roadway, and large gates. The piers were placed to an accuracy of 300 mm or better in plan.

Caissons have been widely used to form submerged tube road tunnels across tidal estuaries. Ref. 1977(4) summarises the experience of a Netherlands company in this field since the beginning of the century. A recent example in the UK is the A55 road crossing of the Conway estuary on the north coast of Wales. This involves placing the caissons in a mean tidal range of 5·2 m; i.e. a range that is much larger than at previous sites of road crossings of this type and one which is of definite interest as regards tidal power. At the time of writing, the caisson units are under construction in an adjacent workyard and some dredging has been done.

One caisson has been constructed and placed in a spring tide range exceeding 10 m. This is the intake caisson for Aberthaw power station, located about 1 km off the north shore of the Severn estuary west of the proposed site of the Severn barrage. The caisson comprised a double skin concrete cylinder, 29 m diameter and 20 m high, founded on the seabed at a level of −13 mOD (metres above Ordnance Datum). At float-out it weighed about 3900 t (Fig. 4.4). The

Fig. 4.4 Intake caisson for Aberthaw power station
(Photo: Edmund Nuttall & Co. Ltd.)

foundations comprised three 7·6 m diameter pads placed on gravel within temporary concrete rings and then undergrouted. The caisson's behaviour when floating and when under tow were model tested at 1:35 scale. The difficulties encountered during towing and placing are described graphically in Ref. 1962(1). Vortex shedding behind the caisson, due to the strong tidal currents, prevented the use of a barge for placing the concrete infill needed to achieve full stability, and so three small helicopters made 20 000 flights to transport nearly 4000 t of concrete as well as men and other materials. This points to the need to avoid underestimating the difficulties of placing and founding large caissons in tidal ranges that are attractive for tidal power.

For the development of the North Sea oilfields, concrete oil production platforms have been built in special workyards adjacent to deep water and, when complete, towed to their final positions in the North Sea. The largest of these, built at Stavanger in Norway, weighed about 1 Mt and is the largest single object moved by man.

These examples show that a great deal of practical experience has been gained around the world with the construction and placing of large precast concrete structures. However, with a few exceptions, the neap tidal range at the sites in question has been relatively small, perhaps 2·5 m or less, so that the time available for placing a caisson at slack water has been reasonably long, and the drag on the caissons during the rise and fall of the tide has been small. By definition, tidal power schemes are located where the tidal range is large, so that towing and placing large caissons will involve large forces and short slack water periods.

4.3 Caissons for tidal power: design

Much work specific to the Severn barrage has been carried out, by the Severn Tidal Power Group in the period 1984 to 1989. The results of initial studies were summarised in Ref. 1987(16). More details are given in the Group's full report (Ref. 1986(17)).

Fig. 4.5 shows a cutaway view of the type of turbine caisson proposed for the Severn barrage as the result of studies carried out for the 1981 Severn Barrage Committee. Ref. 1981(15) describes the basis of design, while important work by Sir Robert McAlpine & Sons Ltd. and by Taylor Woodrow Ltd. is reported in Refs. 1980(9) and 1980(10), respectively. The main factors on which the design was based were as follows.

Firstly, a cellular structure was adopted, a typical cell being a cube with a side 5 m square and walls 300–600 mm thick, depending on location, water pressures and so forth. The interaction of the walls and floors of the cells in three dimensions gives the overall structure good stiffness and allows the various loads to be transmitted efficiently into the structure. The main alternative would be to use cylindrical forms, because a cylinder subjected to uniform external pressure goes into compression and thus reduces the risks of the concrete cracking, but the complications of the space required for the turbines and the intersections of the cylinders mean that the design becomes

Fig. 4.5 Cutaway view of turbine caisson for Severn barrage
(Photo: Department of Energy and Binnie & Partners)

complex. Ref. 1980(10) describes work done to develop a workable design based
on concrete cylinders, a concept which has not been adopted.

Secondly, each caisson should be designed to accommodate at least three
turbines, both to assist the floating stability of the caisson, and to help prevent
the tendency for the caisson to float after it is in position and one water passage
has to be dewatered for inspection or maintenance.

Thirdly, the possibility of accidents should be allowed for. Thus the roadway,
which would be added after the caisson had been placed and which would also
carry the power cables, should be above the level of ship impact, and there
should not be any connections between caissons below high tide level in case
one turbine area is flooded. In addition, ballast chambers at each end, which
could be used to trim the caisson during the tow and during placement, could
be filled with sand, suitably protected against waves, and thus also provide a
cushion against ship impact as well as contributing to the overall stability of the
barrage.

Lastly, a suitable gantry crane or cranes should be available to handle and
place the large gates needed to close off a turbine passage for maintenance or in
the event of a turbine failing to operate properly and 'running away'. The
gantry rails would have to be laid after the appropriate caissons had been set in
position so that any variations in line or level could be corrected. This
presupposes that the turbines have adjustable distributors and no permanent
downstream gates, this subject being discussed in Chapter 3.

The concept of prefabricating structures and floating them into position applies also to the sluices, the design of which is discussed in Chapter 5. Because the sluices are less complex structures and are generally somewhat smaller and lighter than the turbine caissons for a particular barrage, they present rather lesser engineering problems. Thus the rest of this chapter concentrates on turbine caissons.

4.4 Wave forces

Since the upper parts of turbines and sluice caissons facing the sea are likely to be flat surfaces, wave forces will be important in the design of the outer walls. In deep water, waves will normally reflect off a vertical wall. If the timing is right, perfect reflection will result in 'clapotis' or standing waves forming whose height is much higher than the incoming wave. Fig. 4.6 sets out a simplified method of assessing the resulting forces and examples are given in Table 4.1. If the water is shallow enough to cause the largest waves to break, a process which may be assisted by the waves crossing the fast-moving water from the turbines, then a large wave could break against the wall of the caissons and 'slam', generating large instantaneous forces. Fig. 4.7 illustrates this and a method of estimating the force. Again, examples are given in Table 4.2. Both these methods of estimating wave forces are simplified; the subject is complex and appropriate model tests would have to be carried out at the detailed design stage for a tidal barrage located in an exposed estuary.

4.5 Construction

The design shown in Fig. 4.5 would weigh over 90 000 t at float out and so would have a draught of about 22 m. Other features are included in Table 4.3. This draught is about twice that available at any shipyard, and the minimum dimension in plan of about 60 m is greater than that of any normal dock entrance. Consequently, there is no possibility of the concrete caissons to house the turbines for a large tidal power scheme being built in an existing shipyard or dock, and therefore special facilities would have to be built. These could comprise workyards arranged so that the lower parts of the caissons would be built in the dry, behind large caisson gates. When ready, each caisson would be floated out and sunk onto a prepared berth which would allow the upper parts to be completed, but where there would be enough water at high water of spring tides to allow the caisson to be refloated for towing to the barrage. If necessary, construction of the upper parts could be carried out in two stages.

By having more than one workyard available, it should be possible to avoid the situation of having too many people concentrated in one place, and introduce an element of competition, especially if the normal complex bonus system operated on civil engineering sites were replaced by a simple bonus based on the completion of each caisson. Ideally, the promoter of the barrage should be able to offer to pay for the caissons as they are delivered, irrespective of where they were built, until the barrage is complete.

The cost of three workyards required for the Severn barrage has been estimated at over 10% of the cost of the entire project (Ref. 1987(16)). This represents a major item of capital expenditure, which would be incurred at the beginning of construction and thus have an unduly large effect on cash flow. In addition, unless the workyards could be converted into new ports, which should not be difficult but which could jeopardise the future of existing ports nearby, they would be a large unproductive element of the finished project.

4.6 Steel caissons

One alternative to concrete would be to build the caissons of steel, using existing shipyard facilities. Fig. 4.8 shows an outline design of steel caisson prepared by Binnie & Partners for a preliminary study of a tidal barrage in the

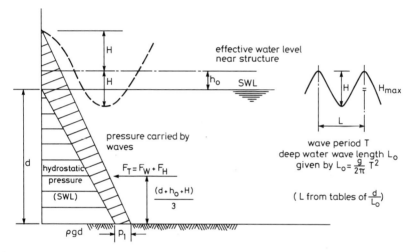

Fig. 4.6 'Clapotis'—i.e. perfect reflection forming 'standing wave'

$$p_1 = \rho g \frac{H}{\cosh(2\pi d/L)} \qquad (1)$$

"Sainflou's method"

$$h_0 = \frac{\pi H^2}{L} \coth(2\pi d/L) \qquad (2)$$

Hydrostatic component of force, $F_H = \frac{1}{2}\rho g d^2$

Total force $F_T = \frac{1}{2}\rho g \left[d + \dfrac{H}{\cosh(2\pi d/L)} \right] \cdot [d + H + h_0]$

Wave force $F_W = F_T - F_H$

Notes
1. Sainflou's method generally gives conservative results.

2. h_0 probably overestimated by using $H = H_{max}$; possibly better to use $H = H_{mean} = \dfrac{H_{max}}{3}$

Table 4.1   Force due to an unbroken wave (Clapotis)
(a) Depth $= 2$ m; $F_H^* = \frac{1}{2}\rho g d^2 = \frac{1}{2} \times 1000 \times 9\cdot8 \times 20^2 = 1960$ kN/m

H(m)	h_0(m)	$T = 8$ s F_T	F_W	h_0	$T = 10$ s F_T	F_W	h_0	$T = 12$ s F_T	F_W
2	0·16	2270	310	0·13	2300	340	0·12	2320	360
4	0·64	2640	680	0·53	2700	740	0·48	2750	790
6	1·44	3060	1100	1·19	3160	1200	1·08	3240	1280
8	2·56	3540	1580	2·11	3690	1730	1·92	3790	1830
10	4·00	4100	2140	3·30	4280	2320	3·00	4420	2460

(b) Depth $= 30$ m; $F_H = \frac{1}{2}\rho g d^2 = \frac{1}{2} \times 1000 \times 9\cdot8 \times 30^2 = 4410$ kN/m

H(m)	h_0	$T = 8$ s F_T	F_W	h_0	$T = 10$ s F_T	F_W	h_0	$T = 12$ s F_T	F_W
2	0·14	4810	400	0·10	4870	460	0·09	4910	500
4	0·54	5260	850	0·42	5380	970	0·35	5460	1050
6	1·22	5770	1360	0·94	5950	1540	0·79	6070	1660
8	2·18	6340	1930	1·66	6570	2160	1·41	6750	2340
10	3·40	6970	2560	2·60	7260	2850	2·20	7480	3070

Notes: Forces in kN/m
F_T = total force
F_W = wave component of force
$*F_H$ = hydrostatic force with still water (for comparison)

entrance to Strangford Lough, south of Belfast (Ref. 1981(3)). This design was based on the use of 7.6 m diameter Straflo turbines similar to that at Annapolis Royal and discussed in Chapter 3, except that the turbine runner was to be mounted on a cantilever axle, unsupported downstream of the runner, in order to simplify installation. The same general concepts of cellular construction and so forth, as developed for concrete caissons, have been applied. The float-out weight of this design would be about 9000 t and so the draught would be only about 2 m. The width of about 40 m could be accommodated in the largest shipyards such as that of Harland & Wolff in Belfast. Once in its final position, the compartments around the water passage would be filled with concrete in order to provide structural strength and weight. The remaining compartments not needed for equipment would be filled either with sand, perhaps oil-impregnated to reduce corrosion, or with concrete. Much more detailed studies have been carried out by Y-ARD Ltd in association with Roxburgh & Partners, summarised in Ref. 1987(12). In order to reduce costs, reliance was placed on rock anchors drilled into the seabed underneath the caissons to provide some of the downward force needed for stability, and thus lighten the design of the caissons themselves. In addition, access to the turbine was limited by a proposal to locate a sluice passage over each turbine waterway.

At first glance, this concept would appear to overcome the problems associated with the construction and towing of concrete caissons discussed above, and also discussed in more detail later. However, three factors offset the balance.

Firstly, because the steel caisson is effectively little more than a skin to contain much the same amount of concrete and sand ballast needed for a concrete caisson, there would be little saving in direct construction costs but perhaps some overall saving because the expensive workyards would not be needed.

Secondly, the prevention of corrosion of the steel structure would need comprehensive and relatively expensive measures, including impressed-current cathodic protection. Ref 1983(3) discusses this aspect in more detail. Particularly difficult would be the protection of the steel in the intertidal zone, where marine life would flourish unless active steps were taken to prevent this. Widespread use of antifouling compounds would probably not be acceptable for environmental reasons. Paint systems can be counted on to provide only short term protection unless sophisticated and expensive materials are used on steel

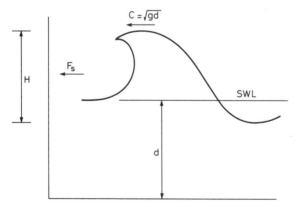

Fig. 4.7 Slamming loads (breaking waves, upper bound solution)

F_S, Slamming force $= \frac{1}{2}\rho C_S . D . C^2$

C_S coefficient in range 2–6 approx.

D "diam. of pile" – assume force acts over whole of breaker height on wall, so use H

C velocity of water $= \sqrt{gd}$ in shallow water or $\dfrac{g . T}{2\pi}$ in deep water

Use $F_S = \frac{1}{2}\rho . C_S . H . (g . d)$

Assume $C_S = 3.5$

$\therefore F_S = \rho . g . d . H$

Table 4.2   Force due to a breaking wave

H	Depth = 20 m (F_H = 1960 kN/m)	30 m (F_H = 4410 kN/m)
2	680	1030
4	1370	2050
6	2050	3080
8	2740*	4120
10	3430	5145

Slamming loads in kN/m

$$\text{e.g. } * \frac{\text{Total force}}{\text{Hydrostatic force}} = \frac{2740 + 1960}{1960} = 2 \cdot 4$$

which has been cleaned and prepared immaculately. Their renewal in the exposed conditions of the sea would be difficult.

Thirdly, shipyards are set up and organised to build ships, and therefore have a wide range of skilled personnel available, such as carpenters, electricians, plumbers and so forth. The simple steel caissons would not involve these skills and so an imbalance in employment would be created. This need not be important if a shipyard is building a one-off steel structure in among its normal work, but the building of a number of steel caissons in a relatively short period to the exclusion of other work could cause problems.

The design modifications discussed earlier introduced some uncertainty into the long term stability and maintenance of the caissons. Taken together with these factors, concrete caissons are considered as yet to be the preferred option. However, the Steel Construction Institute has recently carried out an extensive reappraisal of the design and cost of steel caissons for large, medium and small

Table 4.3   Results of design calculations for turbine caissons (3×9 m dia. bulb turbines)

Dimensions	69·5 m long (cross barrage) × 60 m wide
Weight during tow-out	94 000 t
Draught during tow-out	22 m
Final weight of concrete and ballast	215000 t
Factor of safety: against sliding with one draft tube dewatered	2·31
against overturning with one draft tube dewatered	
against floating with all draft tubes dewatered (construction stage)	1·24
Maximum ground pressure	46 t/m^2

tidal power schemes (Ref. 1989(2)). This has involved the development of a computerised method of optimising the design of the caissons. This study has confirmed that there is little to choose in basic cost between steel and concrete caissons, but the avoidance of the need for purpose-built workyards for concrete caissons could make steel caissons quicker to complete and thus cheaper overall.

4.7 Towing and placing caissons

Many organisations have experience of, or have studied, the problems arising when towing and then placing accurately large caissons in tidal waters. Examples in the literature include Refs. 1962(1), 1974(4), 1979(6), 1980(9), 1983(1), 1984(2) and 1989(2).

Towing and placing the caissons for a large tidal power scheme will present some interesting challenges. These relate to:

- The large tidal range
- The strong currents
- The accurate preparation of the sea bed ready to receive the caissons
- The need to place caissons very close to each other and accurately to line and level

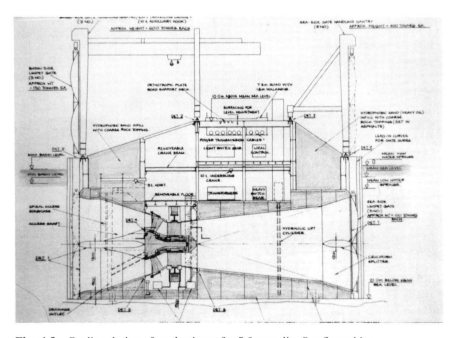

Fig. 4.8 Outline design of steel caisson for 7.6 mm dia. Straflo turbine (Binnie & Partners)

- Sealing the space between the underneath of each caisson and the seabed
- Making sure that high velocity flows during construction do not scour the seabed so as to cause problems in foundation preparation
- The possible effects of long-period, low amplitude swell waves during mooring and ballasting down causing the caissons to move horizontally in an unpredictable manner (e.g. Ref. 1963(2)).

Each of these aspects is worth separate consideration.

4.8 Tidal range

Starting with the construction of the caissons, and assuming that this takes place in one or more purpose-built workyards reasonably near the site of the barrage, then the first challenge is the design and construction of the workyards themselves to accommodate a tidal range which could exceed 10 m. Earlier in this chapter, it was explained that large caissons would probably be built in stages, starting with the base built in a dry dock behind a float-in caisson gate. This base therefore has to be capable of being floated out when the dry dock is flooded, and must have a height which gives a satisfactory freeboard when the base has been sunk on the underwater berm for the next stage of construction. To save cost in the construction of the dry dock, its floor needs to be as high as possible. The maximum possible height will be defined by the draught of the caisson bases floated out on a high spring tide. This would place an undue constraint on the times when the bases could be floated out, and so the floor level could be set at a lower level selected so that the bases could be floated out at high water of mid-range tides. Each would then be moved to the berm for the next stage of construction and ballasted down. The height of the caisson base at the end of the first stage of construction has to be enough for work to be able to proceed throughout the normal maximum spring tide range. When the next stage is complete, the caisson must be capable of being floated off its berm and, preferably, to avoid delay, without having to wait for a high spring tide. Alternatively, all subsequent work after the initial float out could be done with the caisson floating. This would then require a pier or quay which had a water depth alongside at low water of spring tides equal to the finished caisson's draught. This could exceed 20 m. Unless the workyard were located in a well sheltered position, the moored caissons would be subject to movement caused by long-period swell waves. The construction of a quay of this height would certainly be a challenge.

In considering these factors, it has to be borne in mind that transporting men and materials from the shore to one or more caissons moored offshore, as was done for the big North Sea concrete oil platform, would be time consuming and expensive in a situation where rapid, production-line techniques are needed.

If the caissons' late stages of construction have been completed with the caissons sitting on a berm, then the final flotation would probably require a relatively large tide so that the height and cost of the quay are kept to reasonable minima. However, at the barrage, the critical operations of manoeuvering and ballasting the caissons down would be carried out during neap tides so that the slack water periods were as long as possible. Consequently, the caissons would have to be 'parked' for several tides. The parking area would logically be close to the barrage.

4.9 Tidal currents

Chapter 8 discusses the effects of a barrage on tidal currents as work progresses towards completion and 'closure' or, more correctly, the achievement of control over tidal flows. Table 8.2 summarises the changes in the strength of tidal currents predicted for the Severn barrage as construction progresses. These results are for the Severn barrage and are taken from Ref. 1980(5) which refined the results of a previous study based on a flat-estuary model (Ref. 1979(3)). These results will be typical of sites with similar tidal ranges. Also shown are the changes in the length of slack water periods, defined here as the time during which currents are 0·5 m/s or less, or 1·0 m/s or less, and the increases in maximum differential head during the main flood or ebb.

During high water of neap tides, the 0·5 m/s slack water period is halved from 150 min to 75 min at the stage when the last turbine caisson (Stage 3) is placed. Although the practice and experience gained with the early caissons should make the towing and placing crews expert and able to position the caissons efficiently and accurately during the later stages, the logistics of manoeuvering a caisson weighing 80 000 t or more into position with a required accuracy in plan of perhaps 300 mm and ballasting the caisson down while the direction of the current changes will be another major challenge. Any accident, such as a caisson colliding with a caisson already placed with enough force to cause damage, could be very costly in terms of delays to the overall programme.

The speed at which caissons are towed to the barrage site will define the tug power required. Fig. 4.9 shows the approximate relationship between the speed of a rectangular floating box with a cross sectional area of 1200 m² towed through the water and the drag force. This area would be typical for a caisson with a width across the barrage of 60 m and a draught of 20 m. The relationship is approximate because the drag force will be increased where the draught of the caisson is close to the available water depth, owing to the caisson squatting as a result of lower hydrostatic pressure underneath. To generate one tonne of towing force requires about 80 brake-horsepower or 60 kW. The proximity of the tugs to a caisson, and wind, will also affect the power required, but to a lesser extent. What Fig. 4.9 demonstrates is that massive tugs would be required to tow a large caisson at over 4 knots (2.05 m/s). Since this is a typical tidal current in an estuary suitable for tidal power, it follows that towing against the tide would be largely an exercise in standing still. Assuming that the caisson is towed from seaward of the barrage, then if the caisson could be anchored during the ebb tide no power would be required. On the next flood tide the caisson would be transported typically about 15 km by the tide plus the distance through the water achieved by the tugs. This seems a feasible way of saving on tug horsepower and fuel.

Depending on the geometry of the seabed at the barrage site, it may not be possible to tow all the caissons directly to their final positions—the seabed may

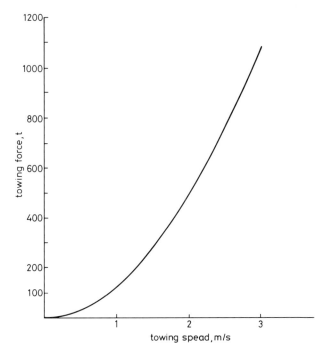

Fig. 4.9   Approximate towing force for caisson 60 m wide × 20 m deep

be too high out of the deep water channel. In this case, the caissons located on each side of the deep water channel, perhaps in an area which would have been dredged, would have to be towed along the line of the barrage and across the main tidal currents. The logistics of this will be complex because an important part of barrage construction will be the minimising of tidal currents through the remaining gaps by allowing the tides to flow through all available openings in caissons that have been placed. Thus the barrage will be relatively transparent to tidal flows and there will be little shelter for caissons being towed along the barrage.

4.10 Foundations

The preparation and subsequent sealing of the foundations for large caissons in an estuary with a large tidal range will present some of the most demanding challenges. The temptation is to throw money at the problem. Fig. 4.10 is an artist's impression of a purpose-built jack-up barge designed to construct a flat concrete bed on which the caissons could be founded. The practicality of this concept is questionable; for example, the newly placed concrete would be susceptible to scour by the turbulent currents, and the enclosed area of the seabed could be expected to trap large quantities of soft sediments in a few tides. Any accident or failure of the device would prejudice the overall construction programme, so at least one spare similar jack-up barge would be required. Having said that, it should be remembered that the construction of the Oosterschelde barrier involved the successful development and use of several large items of specialist floating plant, including a barge equipped to consolidate loose sand deposits, another to lay large areas of special anti-scour mattress, and the catamaran heavy lift barge already mentioned. With this pioneering and successful experience, it would seem sensible to build special plant for a large tidal power scheme to undertake work that would otherwise be difficult or slow if done with normal plant.

In the Severn estuary, the seabed in the area of the Severn barrage consists of a range of materials from hard Carboniferous limestone through marls, mudstones and glacial boulders to soft muds. Other estuaries where barrages have been suggested tend to have a predominance of sands and silty deposits overlying rock. Because the soft deposits are at present in equilibrium, and because the power of currents to erode and transport fine sediments is

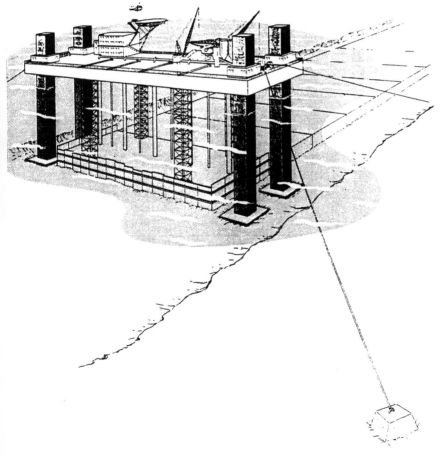

Fig. 4.10 Rig for placing flat concrete base for caissons

proportional to the fourth or even fifth power of their speed, the relatively small increases in tidal currents caused by the start of construction of a barrage can be expected to remove large quantities of soft sediments from the site of the barrage, until equilibrium is regained or the next part of the barrage is placed in position.

If the thicknesses of soft sediments are small, say up to 3 m, then this scouring action could be beneficial in exposing the bedrock at no cost. If sediment thicknesses are large, it may be preferable to design the elements of the barrage to suit a relatively soft foundation, and to prevent wholesale scouring by dredging the seabed to the required levels and then placing a protecting mattress or layer of granular material which has a grading large enough to resist the increasing tidal currents. This is discussed in Chapter 8. Such protective layers would have to be placed before construction of the barrage proper had started. In addition, the barrage would have to be designed to accommodate the appropriate amount of settlement.

Where the foundations of the caissons are rock, excavation will be needed to provide a reasonably smooth and level foundation for the caissons. Rock with a compressive strength of up to about 5 N/mm^2 can be excavated by large cutter suction dredgers equipped with suitable cutter heads, although the maximum depth that existing dredgers can reach is about 30 m, this being dictated by the maximum draught of the largest tankers. For hard rock and/or deeper water, the obvious solution is to use a drill barge to bore holes in the seabed for explosive charges. This method has been widely used for deepening shipping channels and port approaches. However, special equipment will be needed to work effectively in a large tidal range and in strong currents. In addition, the effects of shock waves from blasting on passing migratory fish such as salmon and sea trout can be severe. After blasting, a layer of fragmented rock will be left on the seabed and this will have to be removed. The same applies to rock excavated by cutter suction dredgers, which can leave much debris behind their cutterheads. Consequently, some form of suction dredger would appear to be essential. This could also remove fine silts and other sediments which settle in dredged foundations before the respective caissons have been placed.

An alternative to the complex process of drilling and blasting rock could be to build a purpose-designed vertical boring or grinding machine. Such a device was used in Japan to excavate the foundations for the piers of a bridge. A series of large diameter cutters mounted on a jack-up pontoon were operated on an overlapping pattern to excavate hard rock to a reported accuracy of 100 mm. This method would be appropriate only if the area of rock to be excavated was a large proportion of the total area of foundations, and the rock was reasonably uniform in its properties or, instead, the rock was too hard to dredge and environmental considerations prevented the use of blasting.

Whatever method of excavation is used, the variability of the foundation materials can be expected to result in an uneven finished surface with a certain amount of loose debris. The turbine caissons and sluice caissons would have to be placed to close tolerances in line and level, and so could not simply be lowered onto the 'as dug' seabed. Some form of temporary support would be needed to hold the caisson in its correct position until the space between the underside of the caisson had been sealed. For the Severn barrage, possible options include the construction of large diameter stub piles for each caisson by a special drill barge, or the fitting of each caisson base with a relatively large number of hydraulically extended jacks (Ref. 1890(10)). Other studies have tended to prefer the laying to close tolerances of a granular bed which is then grouted after the caisson has landed. This method is less dependent on the seabed properties being consistent, but the granular material would have to be large enough to prevent its removal by tidal currents, or bound together by some form of fabric. In addition, there will be some risk of high local loads on the underside of the caisson caused by inaccuracies in the surface of the bed material. Since the positions of these load concentrations would be random, the whole of the caisson base would have to be designed accordingly.

Having lowered each caisson onto its foundation, the space underneath would have to be sealed, both to spread loads and to minimise seepage of water underneath when the barrage is operating with large differential heads. The most widely used method is to inject grout made either of cement and water or, if the voids are large, cement, sand and water or even concrete with a plasticising agent to increase its fluidity. If the caisson is supported clear of the seabed, then some form of temporary or permanent seal would be required around the perimeter of the caisson to contain the grout. There is little relevant experience in the conditions of high tidal range and variable seabed applicable to tidal barrages. Inflatable bolsters and semi-flexible skirts have been suggested. Rigid skirts have been used for the concrete platforms in the North Sea, where the platforms have been founded on sediments. This could be appropriate where the caissons are founded on a prepared bed of granular material.

4.11 Long-period waves

During studies for the development of Port Talbot harbour to accommodate ore carriers (Ref. 1963(1)), investigations showed that waves with amplitudes of up to 250 mm and periods generally between 1 and 5 min occurred often in the Bristol Channel. These may be significant if they occur during the critical stages of lowering a caisson accurately into position, because they could cause the caisson to 'range' or move horizontally unexpectedly.

4.12 Post-emplacement work

For several reasons, the caissons housing the turbines and the sluice caissons would not be structurally complete when they were floated into position. The main reasons are that some inaccuracy in placing would be inevitable, so roadways and gantry crane rails would have to be designed to accommodate this, and all parts of the structures above water level would increase the float-out draft. Thus the following parts could be added as precast units, placed by a large floating crane, possibly the heavy lift crane discussed in Chapter 3 as being a method of installing complete turbine/generator units:

- The piers supporting the roadway
- The roadway, in the form of concrete or perhaps steel box girders

- The roof of the turbine caissons, or at least the part over the turbines
- For the sluice caissons, the upper panels which act as wave dissipators and closure panels.

The piers and roadway beams could be built after the appropriate caisson had been installed and the inaccuracy in placement defined. Otherwise, standard units could be used, with their positions in plan being adjustable and relatively small errors in level being corrected by suitable packing.

The electrical and mechanical plant, including gates and their operating gear, would all be installed after each caisson had been placed but in as large units as possible. For example, modules containing lubricating equipment, hydraulically operated control gear, switchgear and so forth could be built and precommissioned, then installed in the same way as they are installed on offshore oil platforms, leaving as little work as possible to be done on the barrage.

4.13 Maintenance

A feature of tidal power is that the predictability of the tides. The spring–neap cycle and the fact that an ebb-generation barrage produces electricity during less than half the tidal cycle all allow routine inspections and maintenance to be undertaken with little or no effect on energy output. Even the loss of say 5% of the turbines for repair results in only about 3% reduction in energy output, because the remaining turbines can be operated to use more water, at least during tides of less than maximum range.

If a barrage were built of caissons, then the ends of the water passages would have to be closed off during float out in order to provide maximum buoyancy, using either large limpet type gates handled by crane or else caisson gates which are floated into position and then sunk by controlled flooding. These gates would remain available to allow water passages to be dewatered for inspection and maintenance. This would be ideal, because the whole of the water passage would be accessible. If, instead, reliance had to be placed on gates located within the water passages, then the outer parts would be inaccessible.

The amount of maintenance that the structural parts of a well designed and built tidal barrage need should be minimal over a long life. Key factors will be the chemistry of the cement and aggregates in the concrete, and the fact that the barrage is unlikely to be subject to abrasion by stones thrown up by waves, as is common in sea defence works. Poorly chosen materials can lead to excessive alkali–silicate reaction, the dreaded 'concrete cancer' which has been receiving much publicity, and to premature corrosion of steel reinforcement. As mentioned at the beginning of this chapter, reinforced concrete structures have survived forty years or more in the open sea even though they were designed for short lives.

Chapter 5
Sluices

5.1 Introduction

Historically, tidal sluices are structures, located in sea walls or river flood banks, which are fitted with gates which allow water from inland drainage systems to flow into the sea or tidal river. In order to avoid having to pump the drainage water from low-lying land, the sluice would be designed to discharge during the lower part of the tide but also to keep out the sea at high tide. Thus the sluice would be fitted with some form of control gate or gates, depending partly on the flow capacity required, which would, if possible, operate automatically.

At one extreme and in its simplest form, a sluice can be fitted with a simple flap gate, the flap gate being made of cast iron, steel, wood or composite materials including plastic. The operation of this type of gate is automatic: the gate opens when the water level behind it is higher than the level outside, and thus maintenance requirements are low. Because of their simplicity, these small sluices rarely fail, and, if they do, the consequence may be little more than some local flooding.

At the other extreme are the large, controllable barriers which control tidal levels in rivers and estuaries. The Thames barrier is one example which has some relevance to tidal power because the cills between the piers were made of reinforced concrete, precast in an adjacent workyard and then floated into position. The largest barrier of this type is the Haringvleit sluices at the mouth of the Rhine in the Netherlands. There are 17 gates each 58·5 m wide. Fig. 5.1 shows a general view. Ref. 1965(1) describes the design and manufacture of the gates and Ref. 1970(1) includes a description of the early stages of construction of the sluices. These sluices are designed to control the level of the Rhine in its lower reaches and thus prevent flooding that would be caused by exceptional river flows or by very high tides, or by a combination of both. In the same way, unduly low river levels are not desirable for navigation, and so, during dry periods, the Haringvleit sluices are used to keep river levels higher than would occur naturally.

The scale of the Haringvleit sluices is comparable with that of the sluices that would be required for a major tidal power barrage. Thus their successful design, construction and operation are of relevance to tidal power.

In theory, a tidal power barrage does not have to have sluices to be able to operate, even if the turbines are designed to operate in only one direction of flow, such as from the basin to the sea. This is because the turbines can be made to freewheel in reverse in order to refill the basin. However, the turbines are not

Fig. 5.1   Haringvleit sluices at mouth of R. Rhine

efficient hydraulically in this mode of operation and, more importantly, the progressive constriction to tidal flows caused by the construction of the barrage will create large differences in tide levels across the barrage, especially during spring tides. This will then create severe problems for the final stages of construction. Consequently, the inclusion of an area of sluices which is relatively large in proportion to the cross-sectional area of the estuary at the barrage site has important benefits during construction as well as during operation. These are discussed in Chapter 8.

5.2 Types of gate

The sluice gates for a tidal barrage will be as large as possible so that minimum obstruction is presented to tidal flows. In addition, they will have to satisfy the following criteria, depending on circumstances and the method of operation of the barrage:

- Open and close reasonably quickly at predetermined times during each tidal cycle, generally when water levels on each side are nearly equal
- Require modest power for operation
- Be efficient in passing large flows at low heads, both during construction and during operation
- Withstand a large differential head of water during the times when the turbines are operating, normally from the basin to the sea but, for two-way generation, in either direction

- Withstand wave loads, as appropriate, from the seaward side and the basin side
- Withstand modest reverse heads; for example, when one gate fails to open for refilling, or when the estuary is at risk from flooding due to a high 'surge' tide and the barrage is to be operated to exclude the top of the tide
- Be resistant to accidental damage that could be caused by mal-operation, by impact by a large floating or submerged object, such as a tree trunk, or by an object preventing full closure
- Not be liable to erosion by suspended sediments or to the deposition of sediments in the corners of gate guides
- Be easy to inspect and simple to maintain
- Be durable
- Have a low first cost

From this list, it follows that sluice gates should have as few working parts as possible located below low water and, to avoid large power inputs during operation, the loads from the gates should be transferred directly to the structures rather than through the operating mechanism.

There are a number of different designs of gate that could be used for the sluices of a tidal barrage. Ref. 1979(2) is a useful review of the subject. Four types of gate have been used or considered seriously for tidal power schemes (Fig. 5.2). These are flap gates, vertical-lift wheeled gates, tainter or radial gates and rising sector gates. Each has its advantages and disadvantages which are discussed below.

5.3 Flap gates

These offer what at first appears to be a major advantage, namely that operation is entirely automatic and no operating machinery is required. In addition, the principal working parts, the hinges, are located at the top of the gate and therefore could be accessible for inspection and maintenance, at least at low tide. Gate design and construction are relatively straightforward. For these reasons, flap gates have been used for tidal sluices on land drainage schemes for centuries, although they are rarely larger than about 3 m square. Gates up to 12 m wide by 8 m high have been used for tide-excluding barriers built on the north bank of the Thames as part of the flood defences seaward of the Thames barrier (Fig. 5.3). In this case, the gates are normally held open and are closed only when an unusually high tide is expected. Because they close under gravity, operation to exclude a very high tide can take place without power being available and is thus fail safe.

Flap gates are susceptible to slamming against their sealing frames when subjected to wave action when they are slightly open. To help prevent this, three solutions are possible. Firstly, the sealing frame can be set at an angle to the vertical, so that a distinct pressure is needed to unseat the gate. Secondly, a positive seating pressure can be provided by having an eccentric pivot instead of a normal hinge. Thirdly, rapid movement can be prevented by mechanical damping, which provides a restraining force which increases as the rate of movement increases. This could comprise either hydraulic rams or splitting the

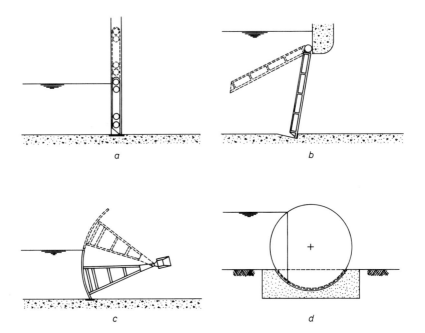

Fig. 5.2   Types of gate suitable for a tidal barrage
a Wheel-type vertical lift gate
b Flap gate
c Radial gate
d Rising sector gate

opening horizontally into two or more openings and linking the gates together so that the top gate is damped by the lower gate(s).

   The feasibility of flap gates larger than about 8 m height is not proven in exposed conditions. A large tidal power scheme such as the Severn barrage could require sluice gates operating in water depths of 20 m or more, so multiple gates would be needed. Fig. 5.4 shows an outline design developed for a tidal power site in north-west Australia where the entrance is narrow and deep so that strong tidal currents have to be accommodated (Ref. 1976(1) also Chapter 12). Fig. 5.5 shows a similar design developed during the Severn tidal power studies (Ref. 1981(1)). This design was based on the concept of a float-in concrete caisson and with the gate frame removable both for maintenance and so that the gate installation could be carried out quickly when the rest of the barrage was complete and ready to operate. Each flap gate was to be 20 m wide by 8 m high, this being judged to be the feasible maximum size. To float the caisson into position, temporary caisson-type gates would be fitted at each end of the water passage.

   This design was model tested at prototype flows up to 2000 m³/s (Ref. 1981(6)), and achieved a coefficient of discharge C_d of about 1, where

$$Q = C_d A (2gH)^{0.5}$$

Fig. 5.3 Large flap gate in raised position
(Photo: Binnie & Partners)

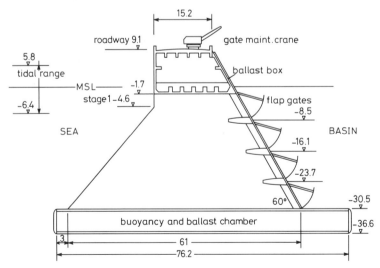

Fig. 5.4 Outline design of flap-gate sluice caisson for deep water

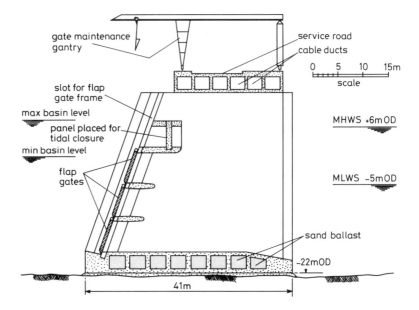

Fig. 5.5 Sluice caisson with flap gates

and Q = flow
   A = area of opening
   H = difference in water level across the sluice.

Fig. 5.6 shows the range of discharge coefficients found, low coefficients resulting at low flows because of the obstructions caused by the part-open gates. At full flow, the gates, which had been weighted to prevent them opening until the prototype head difference was 200 mm, did not lie horizontally but were still inclined downward. This had the detrimental effect of deflecting the flow downwards with the consequent danger of scouring of the seabed.

There should be scope for improving the performance of this design by shaping the flap gates as aerofoils so that they generate upward lift when open by 'flying' in the stream, and thus lie horizontal at large flows.

The inherent simplicity and automatic operation of flap gates are attractive features, but these are offset by the poor discharge performance, the difficulty of preventing slamming under wave attack and by the difficulty of satisfactorily retaining a reverse water pressure.

5.4 Vertical-lift wheeled gate

This is the type of gate chosen at La Rance barrage and is also widely used in hydro-electric projects for emergency closure gates. Fig. 5.7 shows an example of a large gate of this type closing the navigation opening of a tide-excluding barrier located on the river Roding on the north bank of the Thames.

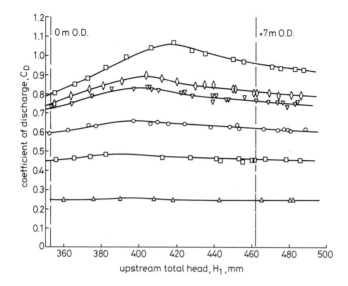

Fig. 5.6 Flap gate sluice: in service: flow from sea ½ model discharge
□ 36·0
◇ 28·1
▽ 25·0
○ 18·6
□ 12·2
△ 5·2

For tidal sluices, the main advantage of vertical lift gates is that they can be fitted into a structure which has a relatively small throat area yet is hydraulically highly efficient. Designs for the sluice water passage have been based on the concept of the venturi tube which has a gently curved inlet and outlet. These accelerate the water flow smoothly up to a maximum at the gate opening, converting potential energy into kinetic energy, and then decelerate the water smoothly in order to recover as much of the kinetic energy as possible. As a result, the gate area needed to achieve a certain flow is kept to a minimum. Fig. 5.8 shows the arrangement at La Rance, where the sluices were designed to work with flow in either direction.

For an ebb generation barrage, the main function of the sluices is to refill the basin to as high a level as possible. Flow in the reverse direction will be important during the construction phase but not after the barrage has been commissioned. The author's experience in water supply engineering lead to the suggestion that a sluice design based on the shape of the Dall tube could have advantages over the classic venturi shape, because the length of the water passage could be reduced without sacrificing any performance, at least in one direction of flow. The Dall tube is widely used in the water industry as a means of generating a pressure difference across a restriction so that flow can be measured, but at the same time recovering a very high proportion of the

Fig. 5.7   Barking barrier, London: vertical lift gate
(Photo: Binnie & Partners)

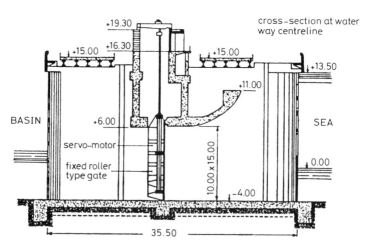

Fig. 5.8   Sluice arrangement at La Rance

difference in pressure in the expansion zone downstream.

Fig. 5.9 shows the outline design of a sluice caisson based on this concept (Ref. 1981(6)) which was model tested (Ref. 1981(6)). Initial results were disappointing, mainly because the flow approaching the sluice from upstream was not symmetrical about the centre line of the opening, unlike inside a pipeline, but had a downward component. Consequently, flow separation occurred at the top upstream edge of the inlet and again downstream of the top of the gate opening. Refinements were introduced, comprising mainly the rounding of the inlet and flattening of the slope of the roof downstream of the gate opening.

Fig. 5.10 shows the 'developed' design, and Fig. 5.11 shows the discharge coefficients achieved, noting that these are in terms of the throat area, and also compares the performance of the two designs and an intermediate 'modified' design. This shows that the discharge performance of the developed sluice was a great improvement on the design first tested. This also demonstrates the value of scale model tests of hydraulic structures where there are novel aspects of the design, because the first design was disappointing even though it was based on a concept which is well proven. The model tests were carried out with a half model (i.e. cut vertically through the centre in the direction of flow, so that the flow pattern could be studied through the glass side of the flume) and at a linear scale of 1:60. This gives the following scale factors:

Velocity   1:7·75
Time   1:7·75
Discharge   1:55 770 (half model to full prototype)

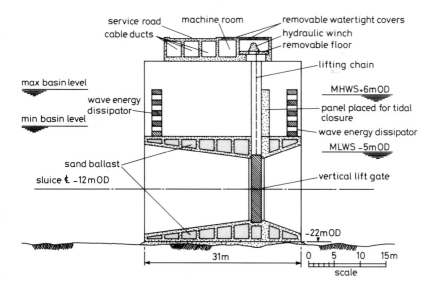

Fig. 5.9   Sluice caisson with vertical lift gate, based on the concept of the Dall tube

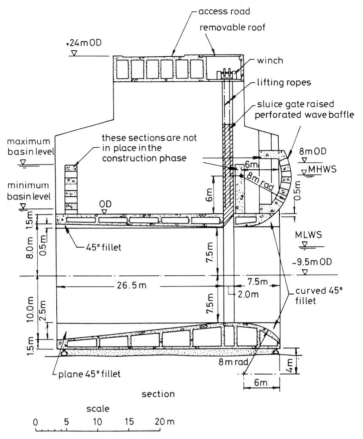

Fig. 5.10   Developed design of sluice caisson with vertical-lift gate

The gate openings in the vertical-lift gate sluice used in studies up until 1986 in the UK have been of various heights from 6 to 20 m but the model tests were carried out on an opening 15 m high. The performance of openings less than 15 m high may be better still because the ratio of the area at the end of the water passage to that at the throat, the expansion ratio, will be greater and so the proportion of kinetic energy recovered should increase. Further model tests would be needed to check this.

   The gate width used so far has been a constant 12 m. This figure was chosen as a balance between the total loads on the gate, and hence gate weight, the proportion of the sluice structure open to flow, and the total weight of the finished structure and hence its stability. Other factors include the unit costs of the various materials involved, the effects on stability of a caisson when one water passage is drained for inspection or maintenance, and the space needed if the gates are to be counterweighted to reduce the size and cost of the operating machinery. The identification of the optimum width of opening in a barrage with a large number of openings is a complex process, and the answer is liable

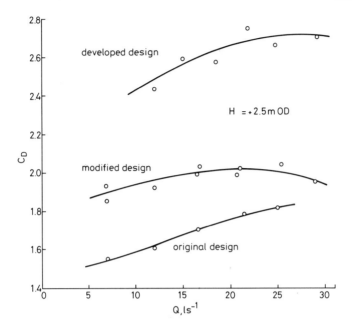

Fig. 5.11 Comparative C_D values: Venturi sluice designs: in service: flow from sea

to change; for example, if steel prices move relative to concrete costs. Table 5.1 summarises the main features and stability criteria for two sizes of gate and three gates per caisson.

In the selection of the height of the gate, the main factor is the level of the seabed relative to the minimum water level at which the sluice will operate, because performance of this type falls off rapidly if the inlet is not submerged. The geometry of the opening shown in Fig. 5.10 effectively defines the level of the bottom of the gate opening at about 4 m above the seabed, or above firm ground. The top of the gate opening should not normally be exposed above low water because, in addition to the drop in performance mentioned above, this will allow waves to run underneath the roof of the water passage and generate large pressure fluctuations when they meet the back of the gate. Much will depend on the wave climate at the barrage site.

Because vertical lift wheeled gates are widely used for emergency closure of water passages through the bottom of dams, their design and performance is well documented. Apart from ensuring the structural integrity of the gate itself, important aspects are the achievement of effective sealing around the edges of the gate, which avoids erosion damage and vibration, and the shape of the bottom edge, which affects the downpull on the gate when it is just open, owing to suction from the high velocity flow. The latter is less important at the low heads of tidal power schemes.

Methods of estimating the weight of vertical-lift wheel gates early in the design process have been developed (Refs. 1986(10), 1986(11)). A knowledge of

Table 5.1   Main features and stability criteria for two sizes of gate and three gates per caisson

Performance of sluice caisson for 12 m square gate

Foundation level	− 20 m OD	− 22 m OD
Total weight of concrete on completion	72,000 t	75,700 t
Approximate weight of gates, counterweights and machinery	750 t	750 t
Factors of safety against sliding (angle of friction 30°):		
with trapezoidal distribution of uplift	1·47	1·38
with rectangular distribution of uplift	1·20	1·13
Factor of safety against overturning (rectangular uplift)	1·53	1·47
Maximum ground pressure (seaward toe)	49 t/m²	55 t/m²
Draught of caisson when under tow	12 m	—

Performance of sluice caisson for 12 × 20 m high gate (three openings per caisson)

Foundation level	− 30 m OD			
Total weight of concrete on completion	105 000 t			
Design conditions:	*Operation*		*1 sluice dewatered*	
basin water level	+ 5 m OD	+ 6 m OD	+ 3 m OD	+ 6 m OD
wave increase	0	3 m	0	0
sea water level	− 6 m OD	− 3 m	− 6 m OD	+ 1 m OD
wave decrease	5	0	0	0
Total differential head	16 m	12 m	9 m	5 m
Factors of safety:				
against sliding (angle of friction 30°)	1·38	1·86	2·06	2·9
against overturning	1·47	1·51	1·47	1·42

steel fabrication costs will then allow costs to be estimated. In Ref. 1986(10) are assembled data for 55 gates ranging in size from 2·7 × 2·15 m high and 2·65 m head, weighing 1 t, to 7·31 × 22·37 m high and 137 m head weighing 343 t. The following relationship was derived for gate weight:

$$G = 0.07 \, KA^{0.93} H^{0.79}$$

where

G = gate weight (t)
A = gate area (m^2)
H = head of water above gate sill (m)

and, for heavy gates ($W^2hH > 2000$ m^4)

$K = 0.9$ for $h/W > 2$
$K = 1$ for $2 \geqslant h/W \geqslant 1$
$K = W/h$ for $h/W < 1$

for light gates

$$K = 1$$

where

W = gate span (m)
h = gate height (m)

As an example, for a gate 12 m span, 6·5 m high and a head of 7·2 m, the formula estimates gate weight at 35·3 t compared with a prototype weight of 27·2 t.

In an earlier article, Erbiste (Ref. 1984(6)) developed the formulas

$G(\mathrm{kN}) = 0.072 \, (W^2hH)^{0.7}$ for heavy gates as defined above and
$G = 0.091 \, (W^2hH)^{0.659}$ for light gates

This provides slightly different estimates. In the example above, the estimated weight is 34·5 t.

To help reduce wave impact loads on the upper part of the sluice structure, perforated walls of reinforced concrete have been included in the outline design. The wall on the seaward side is curved in section to help improve flow patterns in the entrance. Both walls could be precast and could be placed towards the end of construction of the barrage, after completion of 'closure'. In addition, a straight concrete wall has been included just seaward of the gate slot. By leaving this out until a late stage of barrage construction, additional area is made available to water flow during critical closure operations. This wall does not have to extend above wave crest level because the amount of water entering the basin due to waves will normally be small and beneficial as regards energy output.

5.5 Radial gate

Fig. 5.12 shows an outline design of a radial-gate sluice caisson considered for the Severn barrage (Ref. 1981(1)) and Table 5.2 summarises the main engineering features. A gate width of 20 m was selected as being within existing experience and also striking a balance between gate size, pier size and the rigidity of the caisson structure. Gates of this width and 12·4 m height, weighing

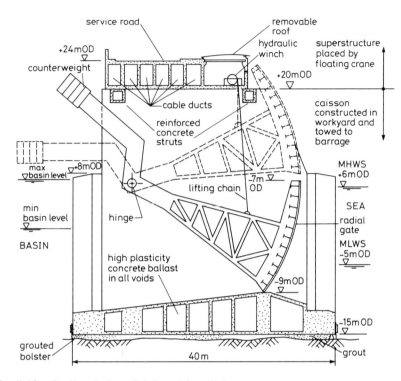

Fig. 5.12   Outline design of sluice with radial gate

Table 5.2   Design conditions and performance of radial-gated sluice caisson

	In operation		*1 sluice dewatered*	
Basin water level	+6	+5	+3 m OD	+6 m OD
wave increase	3 m	0	0	0
Sea level	−3	−6	−6 m OD	+1 m OD
wave decrease	0	3·5	0	0
Total differential level	12 m	14·5 m	9 m	5 m
Factors of safety:				
against sliding (angle of friction 30)	1·4	1·7	2·1	2·6
against overturning	1·7	1·7	1·58	1·48
Total weight of concrete on completion	61 000 t			
Approximate weight of gates, counterweights and machinery	1 000 t			
Draught during tow-out	15·5 m			

75 t, were adopted for the Huntesperrwerk at the confluence of the rivers Hunte and Wesser. They can be closed in 5 min. The gates for the Haringvleit sluices, already mentioned, are 58·5 m wide and the sill level is − 5·5 m. There are two sets of gates, one facing to seaward and the other facing landward. The landward set are higher (10·5 m overall), to exclude waves. This arrangement allows any gate to be raised for maintenance without affecting the overall operation of the sluices. The philosophy here is somewhat different to that appropriate for a tidal power scheme, as the Haringvleit sluices are built for flood prevention and control of river levels.

The radial gate is an attractive design for tidal power schemes for the following reasons:

- Water loads in each direction, both static and dynamic, are transferred to hinges which can be mounted above maximum water level.
- Radial gates can be made in large sizes, and there is much experience in their design, construction and operation.
- With the hinges above water, all moving parts are accessible for inspection and routine maintenance.
- The gate can be counterbalanced to reduce the loads on the operating machinery.
- The gate can be installed and held open ready for operation without affecting water flows during the final stages of barrage construction and 'closure'.
- Migrating fish should not be deterred from passing through an open gate any more than the passage under a normal bridge.

These points would seem to present a strong case for the use of radial gates in the sluices of tidal power schemes. However, radial gates do have certain disadvantages. The first is that radial gates are sensitive to wave forces on their concave faces. The normal design, namely a skin plate supported on horizontal beams which span between the main support members, exacerbates this because the beams form re-entrant pockets which can trap fast-moving water or air and generate high instantaneous pressures. An answer could be to provide a second skin, on the concave side, but this then complicates fabrication, inspection and maintenance.

The second disadvantage is that the large hinge loads are applied to the piers at one end and have to be transferred back into the main part of the structure. This becomes more important as the height of the gate increases. Thirdly, the discharge performance for a given head and per unit area of gate is significantly lower than that of the vertical list gate sluice discussed above.

Refs. 1986(10) and 1984(6), discussed above, also include formulas for estimating the weights of radial gates which have their tops above water. In the first, the following are given:

$$G(t) = K h^{0 \cdot 63} W^{1 \cdot 4} H$$

where $K = 0 \cdot 061$ for $h/W > 0 \cdot 5$

$\qquad K = 0 \cdot 079$ for $h/W \leqslant 0 \cdot 5$

Erbiste (Ref. 1984(6)) gives

$$G(kN) = 0 \cdot 071 \, (W^2 h H)^{0 \cdot 673}$$

As an example, two prototype gates both 15 m wide and about 15·5 m high, for a head of about 15·2 m, weighed 93·8 and 117·7 t, respectively. The estimated weights given by the above formulas are about 102 and 107 t, satisfactorily splitting the two prototype weights.

5.6 Rising sector gate

This type of gate is illustrated in Fig. 5.2. The most famous application is for the gates of the Thames barrier, where the inherent stiffness of the design allowed a 60 m opening to be adopted for navigation, while any failure of a gate to close before a surge tide would not be catastrophic, the remaining gates providing enough obstruction to flows to prevent flooding upstream.

This type of gate was not short-listed during the studies for the Severn Barrage Committee (Ref. 1981(1)), largely because a good seal is required between the gate and the sill to prevent loss of water during the generating period, and this seal would be difficult to maintain. In addition, there was concern that the gap underneath the gate would trap sediments and lead to difficulties in opening the gate. These do not apply to the Thames barrier gates, where there is a 300 mm gap between the gates and their sills. Loss of water through this gap is of no economic consequence.

A slightly modified version of this gate has been proposed for use where the sluice is over the turbine (Ref. 1987(12)). By making one of the longitudinal chambers watertight and buoyant, the forces needed to operate the gate would be reduced. In addition, because the gate is located well above the seabed, there would be little risk of sediment building up in the gap underneath the gate.

5.7 Gate operation

Lewin (Ref. 1980(16)) has estimated that the power required to open a 12 m square vertical-lift wheeled gate in 15 min is about 60 kW, i.e. about 15 kWh. A 20 m wide radial gate lifted 15 m will require about 100 kW. Assuming that similar power would be used to close the gates, which is pessimistic, these represent only around 0.02% of the output of a barrage.

The simplest form of motive power for either vertical-lift wheeled gates or radial gates are chains or ropes passing over sprockets or drums. Each sprocket or drum is driven through a reduction gearbox by a single high speed shaft, itself driven by an hydraulic motor. The motive power for the hydraulic motors for a number of gates could be provided by a single power source. Alternatively, each gate could have its own power source, i.e. a hydraulic pump driven by an electric motor.

The use of oil-hydraulic motors instead of direct electric drive has much to recommend it. Firstly, the hydraulic system is sealed against the corrosive effects of a marine atmosphere. Secondly, protection against overload, especially at the end of travel, can be provided simply and reliably by pressure relief valves instead of limit switches or other mechanisms which rely on electrical contacts which can corrode. Thirdly, safety features such as oil-immersed disc brakes, which cannot be released until the system oil pressure has reached a certain minimum value, can be readily incorporated into an oil-hydraulic system.

An oil-hydraulic system will be more expensive than an electrical system, but, properly designed, will be far more reliable and have a much longer working life.

Each of the radial gates of the Haringvleit sluices is raised and lowered by a pair of hydraulic rams, one at each end of the gate, operating through a mechanical linkage. This removes problems of corrosion of lifting chains or ropes at the expense of requiring a sophisticated system to measure and adjust the travel of each ram so that the gate is not twisted.

5.8 Conclusions

The choice of type of sluice is, in practice, limited to the types discussed above, and will depend on site-specification factors as well as factors such as first cost. For sites with deep water, say 20 m or so at mean tide, the vertical lift wheeled gate appears to be the best choice. For shallower sites, especially those on the routes of migratory fish, the radial gated sluice is probably the best choice.

Chapter 6
Embankments and plain caissons

6.1 Introduction

Only in the exceptional circumstances of a narrow, steep-sided site for a barrage would the 'working' components of a barrage, namely the turbines in their power house, the sluices and the ship lock, occupy the full width of the estuary. The remaining gaps would have to be closed by embankments or plain, i.e. non-working, caissons. This chapter considers both options. Except where access from one bank of the estuary has to be provided early in the construction programme, the non-working parts are the least cost items and should be least sensitive to tidal currents during construction. Thus they would normally be built when the expensive turbine caissons and sluices had been placed. This aspect, loosely referred to as the 'closure' of the estuary, is discussed in Chapter 8.

6.2 Embankments

When considering embankments, the main questions to be addressed are:

- What has been the experience elsewhere?
- What materials would be used?
- How would these materials be transported and placed?
- What are the costs?

Each of these is addressed in the following Sections.

6.2.1 Experience elsewhere
The tidal barrage at La Rance includes a short length of embankment, but this was built in the dry within the area enclosed by the main cofferdams and so was more akin to a normal dam than an embankment built in the sea. The 20 MW pilot plant at Annapolis Royal in Nova Scotia was built in an existing island and so no new embankments were required. Thus there is no direct experience of building an embankment for a tidal power scheme. However, there are several projects where relevant experience was gained.

In Hong Kong in the mid 1960s a dam was built across the mouth of an inlet, Plover Cove, in order to form a fresh water reservoir. When complete, the water behind the dam was pumped out and the reservoir filled with fresh water. The dam was built mainly of dredged sand on a foundation of relatively soft seabed deposits. Fig. 6.1 shows the completed dam, which bears quite a good

Fig. 6.1 Plover Cove dam, Hong Kong
(Photo: Binnie & Partners)

resemblance to an artist's impression of the proposed embankment for the Severn barrage as seen from the Welsh shore at low water (Fig. 6.2). Ref. 1965(3) describes the project in detail.

In the mid 1970s, an even more imaginative project was built in Hong Kong to help meet the increasing demand for water. This was the High Island reservoir and comprised the building of a large dam at each end of High Island to link the island to the mainland. The area enclosed was then converted to a fresh water reservoir with a top water level some 20 m above sea level. What made this project particularly interesting was the building of the cofferdam at the eastern end, in water about 30 m deep, in a location exposed to typhoon waves from the South China Sea. Fig. 6.3 shows the seaward face of the cofferdam, which formed a permanent part of the complete structure.

In the Netherlands, the Delta project has recently been completed. Apart from the Osterschelde storm surge barrier, discussed in Chapter 4, and the Haringvleit sluices (Chapter 5), this project has comprised a succession of large embankments built to close off estuaries or embayments from the sea in order to prevent a recurrence of the disastrous flooding which occurred on 31 January 1953 as a result of an exceptionally high surge tide. These embankments have generally been based on an initial closure structure which has allowed control to be achieved over the tidal flows (Refs. 1965(1), 1972(1)). After closure, the embankments have been completed by covering the initial structure with large quantities of dredged sand with shallow slopes suitably protected against wave

Fig. 6.2 Artist's impression of embankment for Severn barrage
(Photo: Department of Energy)

attack. The experience gained has left the Dutch dredging contractors pre-eminent in their field.

A variety of closure methods was used in the Delta project, including: floating in a series of concrete caissons; a series of caissons with temporary openings which were opened as soon as each caisson had been floated into position, and then all closed at the same time to achieve total closure; and the dropping of thousands of 2 t concrete blocks by a specially built cable car system, thus building up a long weir which eventually appeared above the sea and allowed dredging to begin. A similar system was erected to enable the Osterschelde to be closed from the sea, but this was removed when the environmental lobby persuaded the government that an open barrier with gates should be built instead.

In Germany, the river Eider was closed by a clever system which started with a series of steel frames across the gap. Perforated steel piles were then installed progressively by sliding them down slots in the steel frames so that they spanned 'on edge', thus presenting a moderate obstruction to the tidal flows (Fig. 6.4). Dredged sand was pumped into position between the frames and settled out between the piles, some being carried through the perforations. Additional piles were inserted as required to prevent undue loss of sand, but at the same time to avoid large loads on the frames. The only materials to be brought to the site were the steel frames and piles. All other methods have involved large quantities of rock, concrete blocks or concrete caissons.

The projects described above are important for tidal power in that they demonstrate a variety of engineering skills that could be used to construct embankments for a tidal power scheme. However, they have all been built in

Fig. 6.3   High Island reservoir, Hong Kong: East Sea Cofferdam
(Photo: Binnie & Partners)

locations where the tidal range does not exceed 3·5 m, one third of the typical
spring tide range of a good tidal power site. Consequently, the differential heads
across the structures, and the velocities through remaining gaps, would have
been much less than those that could occur during the final stages of building a
tidal barrage. Other problems follow. For example, temporary (and permanent) protection against waves during construction has to be much more
extensive in area if the still water level can rise and fall over 10 m instead of
3 m.

In 1975, towards the end of a major study of the feasibility of storing fresh
water in a reservoir formed by enclosing a large area of the intertidal foreshore
of the Wash bay on the east coast of England, by an embankment built mainly
of dredged sand, a trial embankment was built 4·5 km offshore (Fig. 6.5). This
trial, costing £2·4 million, was necessary to evaluate the practicalities of
handling and placing large quantities of sand, gravel and rock in a spring tide
range of over 6 m, i.e. about double the range experienced elsewhere. Ref.
1976(5) is a detailed report. This trial highlighted the importance of providing
suitable protection of the dredged sand against scour by tidal currents and
waves if large losses of materials or double handling were to be avoided. It also
demonstrated the feasibility of transporting slope protection materials by sea,
and of handling and placing them using floating plant. Fig. 6.6 shows a 2000 t
capacity coaster being unloaded by floating cranes onto 500 t capacity barges,
this operation being necessary because the water depth at the site was

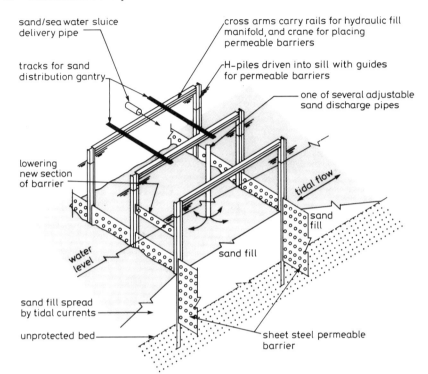

sand/sea water sluice
delivery pipe

cross arms carry rails for hydraulic fill
manifold, and crane for placing
permeable barriers

tracks for sand
distribution gantry

H-piles driven into sill with guides
for permeable barriers

one of several adjustable
sand discharge pipes

lowering
new section
of barrier

tidal flow

sand
fill

water
level

sand fill

sand fill spread
by tidal currents

unprotected bed

sheet steel permeable
barrier

Fig. 6.4 Method of closure used for the Eider dam

insufficient for the coasters. Fig. 6.7 shows the placing of filter material on the sand. The sand was excavated at a rate exceeding 1 t/s by a modest-sized cutter suction dredger from a nearby borrow area and pumped directly to the site via a 700 mm diameter pipeline laid on the seabed.

There are many other marine engineering projects which have been successfully built and where the experience is of some relevance to the design and construction of embankments for a tidal barrage. To take one example, islands have been built of dredged sand in the Beaufort Sea, north of Canada, for oil exploration. These have been located in a modest tidal range but in an exceptionally severe environment as regards ice and wave attack, and weather windows for construction.

6.2.2 Materials for embankments

The material that is most widely available in estuaries is sand. In the Severn estuary, there could be as much as 1000 Mt of sand. In addition, dredgers have been developed which are capable of excavating sand and pumping it up to about 5 km at high rates of production, i.e. 5000 m^3/h or more. Greater distances can be achieved if intermediate booster pumping stations are used, without increasing costs dramatically. It follows that the embankments for a tidal barrage should be designed around the use of dredged sand fill if at all possible.

Fig. 6.5 Offshore Trial Bank, Wash: Completing slope protection

Fig. 6.6 Offshore Trial Bank, Wash: Transhipment of rock

Fig. 6.7   Placing gravel filter on sand

If sand is deposited from the end of a dredge pipe in still water, it will spread until it reaches a stable slope. This could be 1:6 or so, depending on the particle size and variability of the sand. If the dredged sand is exposed to wave action or to tidal currents, then the resulting slope will be much flatter, perhaps 1:50. If the embankment is 20 m high, this would require a base width of around 2000 m. This is not normally practical nor economic, so methods are needed to steepen the slope and thus reduce the finished cross section. A full discussion is given in Ref. 1979(1), the report on embankments prepared for the Severn Barrage Committee. This report remains a useful reference on the subject and so only the main points are summarised here. A novel, alternative method of providing temporary protection of the sand core of an embankment is described in Ref. 1988(5). This involves the use of precast planks inserted into sloping frames. Studies are continuing of this concept, for which there is little relevant experience.

Fig. 6.8 shows a cross section of a 'normal' embankment designed for construction across part of an estuary with a large tidal range, with the embankment being built after the main elements of the barrage have been built and thus subject to the most severe tidal currents.

The key element of the design in Fig. 6.8 is a 'control structure' of quarried rock. This is designed to be built ahead of the main body of the embankment and, as its name implies, to achieve control over the tidal flows. Thus the top of the control structure need not be higher than high water of spring tides. Once above high water, shelter is provided on the basin side for the placing of sand fill and armouring against waves.

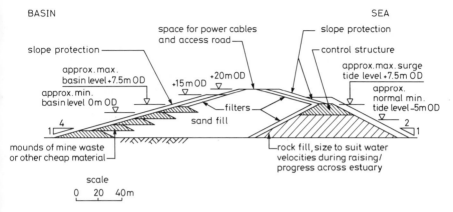

Fig. 6.8 Cross section of embankment proposed for the Severn barrage

The size of rock required for the control structure will vary throughout the construction period. The first material will be placed on the seabed and will offer little resistance to tidal currents. Consequently, the range of sizes of the individual pieces, the 'grading', would be chosen so that material would not be swept away, nor sink into the seabed. If the seabed locally is prone to scour, then it may be necessary to lay first an anti-scour blanket consisting of blocks of concrete tied to a strong fabric, or perhaps a fabric which is rolled out and immediately covered with the first layer of rock. Below low water level, the latter would be the more difficult, because of the difficulty of controlling large areas of fabric in strong tidal currents. The former would be relatively expensive. Such anti-scour protection could be required to be placed over much of the length of the embankment for a tidal barrage before much other progress could be made, simply because the small increases in tidal currents caused by the start of construction could cause wholesale erosion of soft sediments as the tides try to re-establish the cross section of the estuary.

Having placed such an initial layer as the site requires, the main body of the control structure would follow. This would present a fascinating problem of control of the process of selecting the rock at the quarry and transporting it to site, because every load would or should comprise material suited to a particular degree of exposure. Thus, loads placed in the middle of the control structure would have to remain stable only until they had been covered by successive loads. Material placed during neap tides could have a much smaller grading than material placed during spring tides. In addition, the gradings would increase steadily as the project progressed and the estuary became obstructed.

Rock placed on the outside of the control structure would have to be stable not only against the tidal currents and eddies occurring when it was placed, but would also have to resist wave attack until the main wave armouring had been placed. The risk of large waves would depend on the exposure at that part of the embankment, especially whether it was the seaward face or the basin face, on the time that would elapse before the main armouring were placed, and on the time of year. Weather forecasts would be important, particularly if the time before the main armouring is placed were long.

Because rock fill would be at least ten times more expensive than sand, the amount used should be kept to a minimum. Thus the side slopes of the control structure would be as steep as possible. In still water, a slope of 1:1.5 could be achieved. In moving water, a slope of 1:2 would be reasonable, and this is shown in Fig. 6.8.

In the shelter provided by the control structure, conditions would be favourable for the placing of dredged sand. However, the sand placed against the control structure would pass into the rockfill unless a filter layer were placed between the two. In fact, because of the disparity between the gradings of the sand and rockfill, at least two filter layers would be required, as shown in Fig. 6.8. The gradings of a selection of marine sands in estuaries with large tidal ranges are shown in Fig. 6.9. The resulting gradings of the filter layers, and the gradings of known sources of marine gravel are shown in Fig. 6.10.

On the basin side of the sandfill, a very flat slope could be expected, for the reasons mentioned earlier, unless some positive method of retaining the sandfill were adopted. The classic method, widely used in the construction of tailings dams where the tailings from mining operations are deposited in lagoons which are progressively raised, is the 'Christmas tree' system of bunds, as shown in Fig. 6.8. With this system, a relatively small bund of granular material is placed first, in a marine operation by bottom-dump barge, along the toe of the embankment. This material could be of any rock or quarry waste which has material in it large enough to resist the currents and, in the upper layers, the waves expected at that site. Compared with the control structure, which should have been built well ahead of the sand fill, the conditions of exposure of the bunds for the Christmas tree should be much less onerous and so the material would be much smaller and cheaper. Mine wastes and other possible sources of cheap construction materials are considered in Ref. 1974(1).

After the first bund had been placed, dredged sand would be placed until it had reached the level of the top of the bund. The next bund would then be placed and would be founded part on the previous bund, part on the sand fill, thus giving rise to the Christmas tree shape. Again, permanent and more substantial wave armouring would follow as close as practical to minimise the risk of storm damage to the bunds.

One aspect which could be significant is the potential liquefaction of the sand fill by the sudden movements caused by an earthquake. Here the degree of consolidation of the sandfill as it is placed, and the roughness of the individual grains of sand, will be important. The risk of seismic events in the UK is assessed in Ref. 1976(3), where it is shown that a horizontal acceleration of at least 5% g should be allowed for.

The crest of the embankment would serve two purposes. Firstly, it would be the main access route to the working parts of the barrage, both during the later stages of commissioning and during operation and maintenance. Secondly, the high voltage power cables would be laid along the crest. Overhead cables would be at too much risk of corrosion for these to be used for a large barrage. Neither the cables nor the access road should be subject to more than spray during storms, and so the crest level should be set high enough that 'green' water from

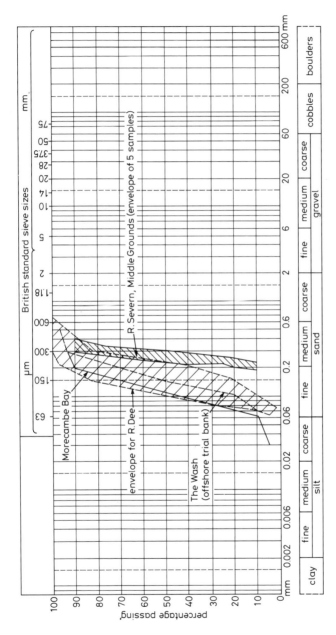

Fig. 6.9 Typical grading curves for seabed sand

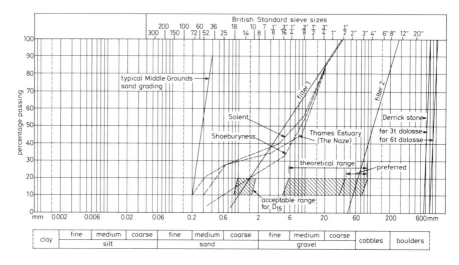

Fig. 6.10   Filter gradings for Severn Middle Grounds sand

large waves does not reach it. This height is a function of wave height, the roughness and the steepness of the wave armouring.

On the seaward side of a large barrage in a reasonable exposed location, such as the Severn barrage, individual waves 10 m high could be expected at high water. With a hydraulically rough armouring system (for this wave height and a slope of 1:2, precast concrete blocks weighing about 7 t would be suitable), the crest of the wave could reach a height of about 15 m above still water level. Thus the crest level would have to be about 20 m above mean sea level; even higher if a smooth armour system were used.

For the sizing of the main armouring system, the Hudson formula gives a useful first guide. This is

$$W = \frac{\gamma_s H d^3}{K_d (\gamma_s - 1)^3 \cot \theta}$$

where

W = block weight (t)

γ_s = density of block material (t/m^3)

H_d = design wave height (m)

k_d = stability factor for that shape of block (from model tests)

θ = angle of embankment slope

It is noteworthy that the weight of the block is proportional to the cube of the wave height.

The value of K_d has to be chosen with care; there have been some spectacular failures of breakwaters protected by precast concrete blocks of special shape.

Fig. 6.11 summarises the sizes of rock or Dolos (a precast concrete block in the shape of an H with one vertical leg turned through 90°; the shape was developed from that of a ship's anchor) for a range of wave heights and degrees of damage. The Dolos sizes are based on $K_d = 10$, a conservative figure. Some damage is considered acceptable during very rare storms.

As an example of the relative importance of the different materials in the embankment design shown in Fig. 6.8, Fig. 6.12 shows the quantities needed for an embankment with a crest height of $+15$ mOD designed for a maximum tidal range of 13 m and maximum wave heights of 8 m and 7 m on the seaward and landward faces, respectively. The lower part of the diagram is enlarged in Fig. 6.13 for clarity. The importance of rockfill in the control structure is clear.

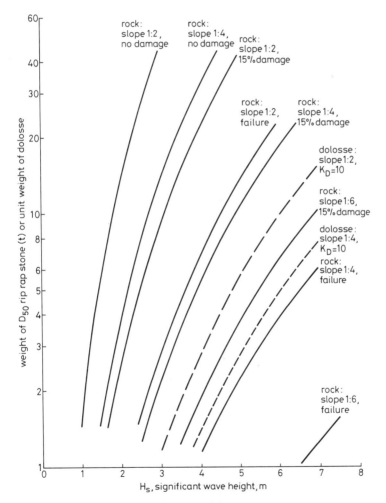

Fig. 6.11   Weights of slope protection material

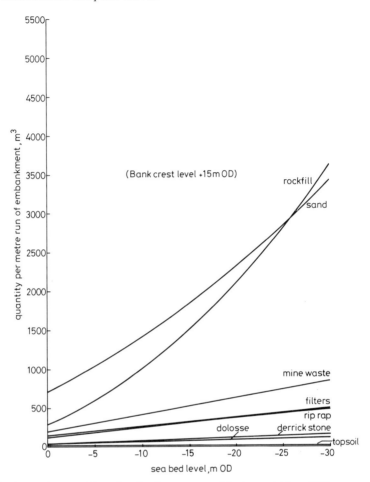

Fig. 6.12 Unit quantities of embankment materials (crest + 15 m OD)
Main embankment
rest level + 15 m OD   min. basin level–6 mOD
crest width 20 m   sea wave Hs 4 m
max. tide level + 7 m OD   basin wave Hs 3·5 m
min. tide level − 6 m OD

6.2.3 Transport of materials
Sand is likely to be readily available within 5 or 10 km of the site of the barrage, and cutter suction dredgers can easily manage this sort of distance, with booster stations as necessary. For the Severn barrage, the total quantity of sand required for embankments would be perhaps 20×10^6 m^3. A single large dredger could excavate and pump this amount in a year, so the production of sand should not constrain overall progress. If suitable reserves of sand are not available locally, then sand could be imported by hopper suction dredgers.

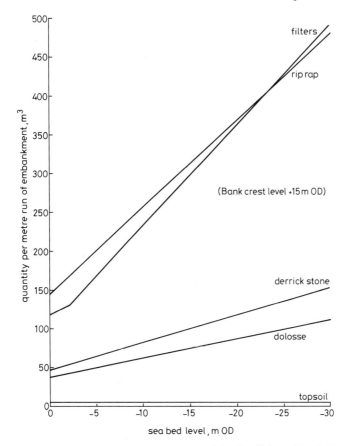

Fig. 6.13   Unit quantities of embankment materials: detail from Fig. 6.12

Modern machines can load up to 4000 m³ in about 20 min, travel to the site and self-discharge through a floating pipeline at the site at much the same rate as a cutter suction dredger. The double handling and greater distance could double the unit cost of the sand, but still leave it a low cost material.

Gravel, if available, would be used for the first filter layer, preventing sand migrating through coarser materials. This would normally be won from licensed seabed deposits and brought to site by a hopper suction dredger.

Costs are discussed in more detail later, but the large quantities of rock fill required for the control structure and for some of the wave armouring, combined with the costs of quarrying, selection and transport, mean that rock will represent most of the total cost of an embankment.

There are two basic methods of transporting rock: overland or by sea. Overland methods are basically road and rail, although canal would have much to recommend it if it were practical at the site in question. To transport large quantities of rock to the site by road would place a severe strain on the local road system and any communities living along the transport routes, especially since much of the rock would be relatively large, i.e. pieces weighing 1 t or more, so that large lorries would be needed, with strengthened bodies. If a new road could be built to each end of the barrage, connecting directly to trunk roads or motorways, then the disruption would be reduced.

Transport by rail would be more logical and environmentally acceptable than road transport; one train can easily transport 500 t of material and special trains with several engines, normally associated with mining, can transport several thousand tonnes at a time. Loading and unloading can be highly mechanised, keeping turn-round times short. The key to overall economy is for the trains to be loaded directly at the quarry. If this is not possible, an obvious reason being that the quarry is not near a railway, then the cost of transporting rock from the quarry to the railhead and the additional handling outweigh any cost advantage that rail transport may have.

If adequate resources of suitable rock are not available close to the barrage site, then transport by sea has much to recommend it. Again, as with rail, the quarry should be as close as possible to the point of loading in order to reduce the number of times the rock is handled. There are quarries which are located on the coast in Scotland, Northern Ireland, Norway and elsewhere. Once a ship or barge is loaded, the cost of actual transport per mile over the sea is small, so that extra distance is relatively unimportant in the overall cost of the material.

Bearing in mind the fact that much of the control structure for the embankment would lie below low tide, the best method of transporting and placing rock for this would be bottom-dump barges or, alternatively, side-dump barges. Such barges can transport 1000 t of rock at a time and, in the case of bottom-dump barges, place their loads in seconds. Side-dump barges push half their loads off one side, half the other, so take a few minutes to discharge unless they can discharge both sides at once.

Bottom-dump barges draw about 4 m of water when fully laden, and have to clear their loads when empty. Thus they need about 5 m of water to work safely. At high water of spring tides, they could dump material onto an embankment which had reached a height up to about mean sea level. To be able to dump material at any slack water, i.e. at high water and low water, they could not be used above about 10 m below mean sea level. Consequently their operation would become rapidly constricted as rock placing approached the level of low water of spring tides. On the other hand, most of the material in the control structure would lie below this level anyway.

Side-dump barges also draw 4 or 5 m of water when laden, but do not need to float over the load after it has been placed. Thus they need less water to work in safely.

When the control structure had reached a height above which barges could not be used safely, rock would have to be placed either by floating cranes or by land-based plant running along the crest of the control structure. Both methods have been used, and the latter is widely used for the closure and diversion of large rivers during the early stages of building dams. Large floating cranes, although effective for the lower parts and essential for the placing of individual concrete armour blocks, would have to work at large radii, and thus reduced capacity and output, to place materials in the upper parts of the embankment.

The operation of large floating cranes on the seaward side of a barrage would be more hazardous than on the basin side, and both sides would become more exciting as the final gap in the control structure narrowed and tidal velocities built up.

Construction of the remainder of the embankment would not wait for the control structure to be complete. The placing of sand fill, filter layers and wave armouring would all proceed logically and as close as reasonably possible behind the advancing control structure. This would help reduce to a minimum the risks of storm damage to the part-complete embankment. In addition, the large area of sand fill above high water would form a useful work area for lorries or off-highway dump trucks hauling rock and wave armour materials to the plant placing these on the upper parts of the embankment.

6.2.4 Costs

The unit costs of the materials to build embankments in the sea tend to be site specific. In assessing these, the factors to be taken into account include:

- the quantities and required rates of production involved,

- the locations of sources of suitable deposits, especially rock of good quality,

- the number of times the material has to be handled, and any constraints on methods of transporting materials to site,

- the risks of bad weather along sea-borne transport routes and at the barrage site, and

- whether port and harbour dues or royalties are payable.

Table 6.1 lists typical unit costs for embankment materials as assessed for notional sites on the east and west coasts of the UK. These are shown to highlight the importance of location of a source of suitable rock relative to the barrage site. The west of the UK has plentiful resources of granite, diorite, limestone and other hard rock. The east of England has few reserves, the nearest source to the Wash for example, being the limestone of the Peak district. To build the trial embankment in the Wash mentioned earlier, it proved economic to import limestone from Belgium.

6.3 Plain caissons

At the transition between an embankment and either sluice caissons or turbine caissons, the end of the embankment will have to be built to a safe slope, and this end will wrap around the first caisson(s). The risk cannot be taken of material from the embankment being swept through the openings of a working caisson. Thus some 'plain' caissons will be essential.

Elsewhere in shallow water, embankments will be the least cost solution. As the water depth increases, the width of the base of the embankment increases, and the unit sizes and layer thicknesses of the wave armouring materials increase because larger waves can reach the embankment. Thus the total cost of an embankment does not increase linearly with water depth. Plain caissons, on the other hand, have to be designed to resist a certain differential head of water

Table 6.1(a) Budget costs of quarried materials for the Liverpool Bay area (1986 prices)

Item	Rip rap		Large filter (200–5 kg)				Quarry waste	
Material	Carboniferous limestone	Granite	Carboniferous limestone	Carboniferous limestone	Granite	Slag	Carboniferous limestone	Carboniferous limestone
See note:		(ii)		(i)	(ii)			(i)
Price ex quarry	9.00	3.60	9.00	7.50	3.60	11.70	5.40	6.00
Loading at quarry	3.00	1.00	1.50	1.50	1.00	1.50	1.00	1.00
Transport	7.00	1.00	7.00	2.50	0.70	7.50	7.00	2.50
Unloading at quay	3.00	1.00	1.50	—	1.00	1.50	1.00	—
Loading into barge at quay	3.00	3.00	1.50	—	1.50	1.50	1.00	—
Transport to site	2.00	2.50	2.00	—	2.50	2.00	2.00	—
Placing on island	12.50	12.50	9.50	9.50	7.50	9.50	6.25	6.25
Total price ($£/m^3$)	39.5	25.20	32.0	21.00	17.80	35.20	23.65	15.75

(i) These figures illustrate the saving which could be achieved if a quarry with a quay could be used.
(ii) Prices if *only* two grades of materials, rip rap and the rest. The condition of these low prices would be that everything would be taken from the quarry.

Table 6.1(b) Budget costs of quarried and waste materials for the Wash (1986 prices)

Item	Rip rap		Large filter (200 cm-5 kg)				Quarry waste	
Material	Granite	Granite	Granite	Granite	Flint	Slag	Granite	Granite
Price ex quarry	12.60	19.35	12.60	14.04	7.20	9.00	5.40	9.90
Loading at quarry	3.00	3.00	3.00	3.00	1.00	3.00	3.00	3.00
Transport	5.95	8.40	5.95	8.40	2.10	6.60	5.95	8.40
Unloading at quay	3.00	3.00	3.00	3.00	1.00	3.00	3.00	3.00
Loading into barge at quay	3.00	3.00	1.50	1.50	1.50	1.50	1.00	1.00
Transport to site	2.00	2.00	2.00	2.00	2.00	2.00	2.00	2.00
Placing on island	12.50	12.50	9.50	9.50	9.50	9.50	6.25	6.25
Total price (£/m³)	42.05	51.25	37.55	41.44	24.30	34.60	26.60	33.55

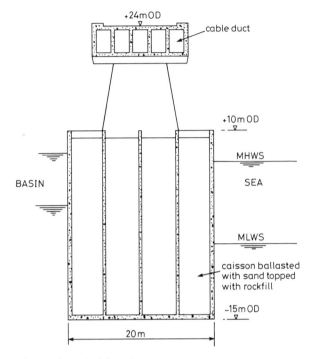

Fig. 6.14 Outline design of plain caisson

and wave impact loads. The former is more important as regards the width of the caisson in deep water. Consequently, the cost of plain caissons does not increase with water depth as fast as that of embankments, and there will be a cross-over point where plain caissons become the cheaper option, assuming, of course, that the seabed is suitable for supporting plain caissons.

The quality of the seabed is much more critical for plain caissons than embankments. As a general rule, the edges of an estuary contain more soft or erodible deposits than the middle part, where the low tide currents are concentrated. Thus there are likely to be much greater problems in arranging foundations for plain caissons near the edges, avoiding:

- erosion and scour of the sediments before the caissons are placed,

- adequate bearing capacity of the sediments,

- leakage under the caissons which, if uncontrolled, could lead to piping and erosion of material at the downstream edge of the caisson, and

- scour of the sediments caused by downwash from waves meeting the vertical face of the caissons, especially where the face of the caisson is formed of cylindrical shapes which concentrate the wave energy.

Fig. 6.14 shows an outline design of a simple plain caisson prepared for the Severn barrage (Ref. 1981(1)). This was shown to be adequately stable during

Table 6.2   Performance of box caisson

Basin water level	$+4$ m OD	$+4$ m OD
Sea level	-6 m OD	-6 m OD
wave decrease	6 m	4 m
Total differential head	16 m	14 m
Foundation level	-25 m OD	-15 m OD
Caisson length (along barrage)	60 m	60 m
Caisson width (across barrage)	25 m	20 m
Draft during tow-out*	15 m	13 m
Weight of concrete in place	20 000 t	12 000 t
Total weight in place	91 000 t	48 000 t
Factors of safety:		
against overturning	1·53	1·63
against sliding	1·25	1·33
Maximum ground pressure	42 t/m^2	31 t/m^2

* Including sand ballast for flotation stability

flotation and during ballasting down, as well as in service. The results are summarised in Table 6.2. A good foundation would be necessary, with a safe bearing strength of about 40 t/m^2. Otherwise, the foundation preparation and support system would be the same as that for the turbine and sluice caissons described in Chapter 4, with the proviso that wave erosion of the seabed along the faces of the plant caisson could be much more severe than for the other caissons. In the same way, the road and its support piers would be added after each caisson had been placed, thus saving on float-out draught and allowing small inaccuracies in placing to be corrected.

The overall conclusions to be drawn from these discussions may be summarised as follows:

- Plain caissons are necessary at the junctions of embankments and 'working' caissons, to provide for the end slope of the embankment.

- Embankments can be, and have been, built on a wide range of seabed types but not in the tidal ranges which are of most interest for tidal power.

- The availability of good quality rock convenient for transport to the barrage site is likely to be a major factor in embankment costs.

- Plain caissons are likely to be cheaper than embankments in mean water depths greater than about 15 m, but the quality of the seabed will be important for the long term stability of the caissons.

Chapter 7

Locks for navigation

7.1 Introduction

If a tidal power scheme is to enclose a part of an estuary where there are one or more ports or quays used by commercial shipping, then either suitable locking facilities will have to be provided or the ports closed down. If the ports are losing money, as is sometimes the case, then it could be argued that there would be economic benefit in closing them, as well as a saving in the cost of the barrage by not having to include locks. However, this raises all sorts of questions about employment, knock-on effects and so forth which are outside the scope of this book, and so the assumption is made that adequate provision would have to be made to allow shipping to continue.

7.2 Ship operation in tidal estuaries

In estuaries with large tidal ranges, ship operators have to plan the movement of their ships with careful regard to the tides, and the care that has to be taken increases as the ship size and the tidal range increase. The ship will have to berth either against a quay which allows for the rise and fall of the tide, or in a dock which has a relatively constant water level, and so the ship has to pass through a lock. In estuaries with large tidal ranges, quays are relatively rare and most ships use docks. If the ship is fully laden, then the clearance between the underside of the ship's keel and the sill of the lock may be so critical that the ship can enter the lock only at high water of spring tides of a particular minimum range. If another ship arrives at the lock on the same tide, then there may not be enough time for both to lock through during one high water, and there can be further complications if one or more ships are ready to lock out at the same time.

A ship travelling up an estuary so as to arrive at the lock shortly before high water will have the benefit of the flood tide current increasing its speed over the ground. Similarly, a ship starting to travel down the estuary at high water will benefit from the ebb tide. In both cases, water depths will be near maximum to assist clearance in any shallow parts of the estuary, although ships will normally be restricted to a well defined and marked, and possibly mobile, deep water channel.

Although ship operators can and do plan the movements of their ships to take into account these factors, it is a common sight to see a number of ships at anchor off the mouth of an estuary, waiting for the flood tide and/or for pilots or or berths. This waiting is unproductive time and is expensive. One effect is that ports located in estuaries with large tidal ranges have to be very competitive to attract custom that would otherwise use ports where the tides are smaller and less troublesome.

7.3 Effects of a barrage on tide levels

As described in Chapter 2, the principal effect of a tidal power barrage operating as an ebb generation scheme will be to raise low water levels in the enclosed basin to around mid tide level. This will remove or reduce restrictions on the movement of ships which would normally not be able to sail at around low water. Associated with this will be the increased practicality of quays (or riparian berths) located along the edges of deep water channels.

The next most important effect will be the time during which high water will remain high, namely around two or three hours instead of perhaps less than half an hour, with an associated reduction in current strengths. This will make it much easier for ships to lock into and out of the entrances at ports, and it may be practical to open the lock gates so that ships could pass through without waiting for the normal operation of the lock.

The third major effect follows from the general reduction in the strength of the tidal currents, and hence their ability to move sediments. This can be expected to reduce dramatically the amount of dredging that port authorities have to do in order to maintain the approaches to their docks. At present, this dredging can amount to several million tonnes a year. The subject of sediment movements is discussed in more detail in Chapters 9 and 10.

Lastly, because a tidal power barrage is extracting energy from the tides, the tidal range to seaward of a major barrage can be expected to be reduced slightly, and this appears to have happened at La Rance. In the case of the Severn barrage, this reduction is predicted to be quite significant at about 10%, i.e. 0·5–1·0 m, depending on the tidal range at the time. Thus, in broad terms, high water levels will be reduced somewhat on both sides of the barrage. Behind the barrage, this reduction is likely to increase, and therefore be progressively more important the further up the estuary a port is located. The significance of this depends on the relative sizes of the ships using a port and the entrance dock. If a particular ship is able, within the constraints of timing relative to spring tides and so forth discussed above, to enter a lock with little under-keel clearance, then the loss of, say, 0·5 m of water depth could mean that that ship can no longer negotiate that lock on that tidal range. If the result is that only relatively rare tides are usable, then the viability of that ship continuing to trade through that port will be at risk. This problem could be mitigated to some extent by the high water stand, because the minimum under-keel clearance set by the port authority for a ship has to take into account the short time of normal high water and so a safety allowance has to be built in, in case the movement of the ship into the lock takes longer than expected. With a prolonged high water, the under-keel clearance could be reduced.

Fig. 7.1 shows the changes in availability of water depth over the range of tides in the Severn estuary for the port of Cardiff as a result of operating the

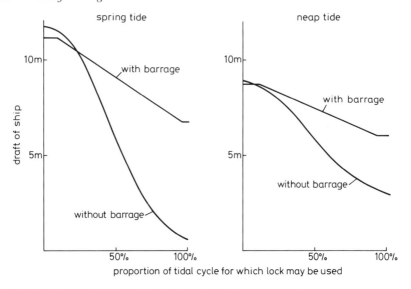

Fig. 7.1   The availability of the lock at Cardiff with and without the barrage

Severn barrage as an ebb generation scheme, based on the results of the 1-D model reported in Ref. 1980(5). This demonstrates the loss of water depth at high tide (in the absence of pumping from the sea into the basin) which would adversely effect the largest ships able to use the lock unless a smaller under-keel clearance were adopted, and the much greater availability of water for ships of modest draught.

Table 7.1 shows how the durations that water levels are above selected levels would be reduced by the Severn barrage for ships travelling to Portbury, near Bristol and the largest dock in the estuary, and to Sharpness towards the top of the estuary. In each case, the top of the existing tide would be lost but there would be water at lower levels for much longer than at present.

7.4 Locks at the barrage

Apart from the question of under-keel clearance discussed above, the other aspects are beneficial to shipping. What has not been discussed so far is the time that ships would take to pass through the locks in the barrage. This is the principal aspect of concern to the ports because of the risk that the apparent delays the ship owners can reasonably expect to occur could sway the decision against using, or continuing to use, that port in favour of another port with less complex navigation requirements.

A study carried out by Rendal, Palmer and Tritton (Ref. 1980(17)) included the results of a computer simulation of shipping movements in the Severn estuary with different arrangements of barrage locks, taking into account tidal range and the random nature of the timing of ships' arrivals at the ports. The simulations showed that a single, very large lock 370 m long by 50 m wide and

Table 7.1 Periods for which water levels are exceeded at Portbury and Sharpness in the Severn estuary

Portbury

Level m OD	Spring tide		Neap tide	
	Existing estuary (h)	With barrage (h)	Existing estuary (h)	With barrage (h)
6·0	1·50			
5·5	2·00	2·40		
5·0	2·45	3·70		
4·0	3·25	5·65		
3·0	4·00	7·60	1·40	
2·5	—	—	2·65	5·85
2·0	4·70	9·30	3·55	6·90
1·0	5·45	11·80	5·10	8·90
0·0	6·15		6·50	12·25

Sharpness

Level m OD	Spring tide		Neap tide	
	Existing estuary (h)	With barrage (h)	Existing estuary (h)	With barrage (h)
7·0	0·70			
6·0	1·70	1·00		
5·5	2·10	2·10		
5·0	2·45	4·05		
4·0	3·25	5·20		
3·0	4·05	7·25	1·95	1·60
2·5	—	—	2·80	6·50
2·0	5·20	9·05	3·65	7·20
1·0	6·60	11·65	5·40	8·90
0·0	8·50		7·75	12·35

able to accommodate the largest foreseeable size of ship able to use the estuary, namely a 150 000 t dwt ore carrier, would cope with up to about 7000 ships a year with an average passage time through the lock of about three hours. With 10 000 ships a year, the average delay rose to an unacceptable 23 hours. These traffic figures compare with the traffic in the estuary at the time of about 5500 ships a year.

If, instead, two locks were provided, the model showed that the average passage time would decrease to about 1·5 hours with traffic of up to 10 000 ships a year. The passage time depended slightly on the size of the second lock, as may be expected, but a lock of only 200 m by 30 m would cause an additional delay of only about 6 minutes unless the traffic rose much above 10 000 ships a year.

The same model was used to predict the changes in the times for ships to lock into and out of the main ports as a result of the barrage. The effect of the long high water stand discussed earlier was to reduce the times by an average of about 1.5 hours. This saving would almost equal the loss of time at the barrage locks. Exceptions would be ports well up the estuary, where the greater loss of depth at high water could result in much more difficulties for the largest ships able to use those locks unless the lock entrances were deepened to compensate. This would be expected to be difficult and expensive, with the added problem that the process of reconstruction would prevent ships using the port and thus lose custom.

The conclusion to be drawn is that, for an estuary used by between 5000 and 12 000 ships a year, two ship locks should be provided at the barrage, one of these being large enough to accommodate the largest ship using, or planned to use, the estuary. This combination would also allow one lock to be closed for inspections or maintenance without causing catastrophic delays. The lock should, of course, be located where there is enough water at low tide for ships to enter and leave safely; otherwise severe delays would be caused as ships queued up to pass through the locks during limited parts of the tidal cycle.

Where the number of ships using an estuary is much greater than about 10 000 a year, such as the Humber estuary where there are about 15 000 ships a year (30 000 movements a year), then either additional locks must be provided or another solution found. One possibility is the provision of one or more navigation openings similar to those of the Thames barrier, i.e. openings 60 m wide closed by rising sector gates. These could be opened during the flood tide, when the basin is refilling, to allow ships up to about 10 000 t dwt unhindered, even accelerated passage into the basin. A separate opening for ships proceeding downstream may be feasible, but the total area of sluices, including the navigation openings, would have to be enough to prevent the speed of the currents through the openings being unsafe.

On the seaward side of the barrage, appropriate breakwaters and guide walls would be needed, depending on the exposure of the site to storms and the proximity and disposition of the turbines. Similar facilities may be required on the basin side, but, if this is sheltered, only guide walls may be necessary. Here the results from suitably detailed 2-D models of water movement will be relevant at the planning stage. These are discussed in Chapter 9. A waiting area for tugs would also have to be provided, together with suitable moveable bridges to allow road traffic across the locks.

Mixing private yachts and large merchant vessels in the same lock is perilous for the small craft. The slightest untoward movement of a large ship will cause severe damage to the yachts, while sailing literature has many instances of yachts being damaged by the wash from the propellers of a large ship carrying out a difficult manouevre in a lock or harbour. The best solution may well be the provision of a small boat lock located in shallow water well away from the main locks. Alternatively, it has been suggested that a boat lift, comprising a tank running on rails over the top of the barrage, may be cheaper. This type of lift has been used on canals in Europe to avoid building a long flight of locks. The change of slope as the tank crossed over the barrage crest and travelled down the other side would be a novelty, not just for the design of the tank but also its operating mechanism, because the canal versions are on a single slope and usually have two tanks linked by a chain or rope passing around a pulley system at the top of the incline so that their weight is balanced and the motive power is minimal.

In order to allow ships to sail to and from the ports during construction of the barrage, the locks would have to be built and commissioned by the time the rest of the barrage had reached the stage where safe passage of ships through the remaining gaps was no longer possible. Just when this stage would be reached will depend on the details of the barrage and, to some extent, on the method of construction. If the parts of the barrage that have been completed are open to tidal flows, the increase in current velocities in the remaining gaps will be less than for 'solid' elements. Also, if construction is based on the use of precast structures floated into position, then the movement of these unwieldy structures will present something of a hazard to shipping. Thus it follows that the construction of the locks is likely to be critical to the overall programme and would have to start as soon as possible.

A ship lock of modest size could conceivably be built as one or more prefabricated units floated into position, in the same way as is favoured for the caissons housing the turbines. A caisson to house four turbines of runner diameter 9 m would be about 60 m wide and 80 m long, so five units of this size could form the basis of a lock 370 m by 50 m, the largest size that would be needed for the Severn barrage. Construction would take place in the same workyards proposed for the turbine caissons, and foundation preparation, towing and placing would involve the same plant and methods. This method should be simpler and more attractive the smaller the size of the lock, and therefore should be particularly appropriate for small barrages where commercial shipping is limited to fishing vessels or where only recreational craft have to be catered for.

The alternative to prefabricating the locks would be to build an artificial island of dredged sand, suitably protected against waves and currents, and then build the lock within the island. This is the method proposed for the Severn barrage in Ref. 1980(17) and in Ref. 1986(9), as well as for the Mersey barrage (Ref. 1988(7)). Compared with a prefabricated lock, it offers the following advantages:

- Work can start as soon as the dredgers, rock-carrying barges, tugs and so forth can be mobilised, and does not have to wait for the first workyard to be completed

- The artificial island will provide a useful working area and, after completion of the lock, not all the island would have to be removed and so the remainder would still provide space for plant or accommodation.

The construction of the lock walls within an island of dredged sand would be done using the diaphragm walls. These comprise the excavation of narrow (0.5–1.2 m) trenches using purpose-built plant, with the sides of the trenches being supported by a slurry of bentonite mixed with water. Bentonite is a special type of clay. When the excavation of a panel of wall has been completed, temporary stop ends are inserted at each end, a cage of reinforcing steel is lowered into the bentonite, and the bentonite is replaced with concrete which is poured down vertical pipes which have their outlets immersed in the wet concrete. This prevents the incoming concrete being contaminated with the bentonite. The bentonite is recovered for reuse. This system is well proven, having been used for a variety of structures requiring deep walls, for example road underpasses and, more relevantly, for the walls or large docks at Bristol and Redcar (Refs. 1973(2), 1973(3)).

The main disadvantage of this concept lies with the construction of the artificial island and, more particularly, with the protection of the sand. Here the same problems arise as are discussed for embankments in Chapter 6, but, in addition, much of the materials protecting the outer slopes would have to be removed to form the entrances to the locks. This can be expected to be difficult, because visibility underwater will be negligible and the various materials will tend to become mixed together as the recovery process gets under way, and the parts of the island on each side of the cuts will still require protection.

A lesser problem that can be foreseen is the transport of men, plant and materials to an exposed site unconnected to the shore until construction of the rest of the barrage is fairly well advanced. This will be costly and weather-dependent.

7.5 Lock gates

The lock gates of a tidal barrage designed for ebb generation or ebb generation plus pumping at high water will have to withstand a large differential head of water between the basin and the sea, typically about two-thirds of the tidal range, and a much smaller differential head in the opposite direction, while the basin is refilling. For a head of water acting in one direction only, the classical choice is the mitre gate (Fig. 7.2) which is simple to build and operate. Ref. 1973(1) describes one large installation of this type. Mitre gates could be designed to withstand a small reverse head, by providing suitable operating gear such as large hydraulic rams acting at the appropriate points. An alternative is a sliding gate which retracts into a slot outside the lock (Fig. 7.3). This type can be designed to withstand heads in either direction, but the space required for the retracted gates is large and expensive to provide in a barrage. It is also more at risk from serious damage due to accidental ship collision or from maloperation.

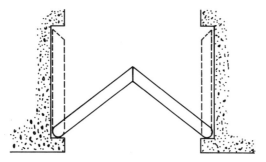

Fig. 7.2 Mitre gate (plan)

For differential heads in either direction, especially for a barrage designed to operate as a two-way generation scheme, the first choice is the vertical axis radial gate. This is the type adopted for the lock at La Rance barrage. Its main advantage lies in the loads from the difference in water level being transferred into the lock walls via the gate hinges, so that the operating gear has only to overcome friction and the drag of the moving gate.

If a ship lock has to be provided which is large enough to accommodate ships which only rarely travel to the ports in the enclosed basin, there will be operational advantages in providing a third set of gates part way along the lock. This will reduce the time needed for water levels in the lock to equalise but, more importantly, will provide a back-up method of allowing small ships to continue to lock through in the event that one set of gates is damaged or requires maintenance.

7.6 Breakwaters

The need for breakwaters to provide sheltered conditions for ships approaching the locks from the seaward side, and possibly from the landward side as well, has been mentioned earlier. For large locks in an exposed estuary, the

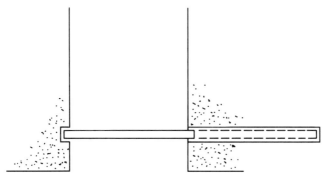

Fig. 7.3 Sliding gate

associated extensive breakwaters would be major civil engineering structures in their own right, involving model tests in wave tanks, detailed geological investigations of the seabed and so forth. So far, only preliminary arrangements and designs have been drawn up in order to arrive at realistic estimates of likely cost. These have been based on the assumption that the breakwaters would be built of a hearting of rock fill, with any face exposed to large waves being protected with precast concrete armour blocks. Compared with embankments for the barrage, there are two areas where savings are possible. Firstly, the breakwater would not have to withstand a differential head of water nor be watertight, so simple rock fill would be adequate. Secondly, there would be no need for traffic to travel on a routine basis along the crest, so the crest could be set much lower than the crest of the main embankment. Indeed, it may be possible to have a crest level only just above maximum tide level, because this would still prevent large waves from reaching the enclosed area, and strong winds will present greater problems for manoeuvering ships than small waves.

The breakwaters could be built of concrete caissons, as was done to create the breakwaters for Brighton marina (Ref. 1979(8)). Unfortunately, vertical faces reflect waves efficiently and so additional rock fill beaches or other methods of dissipating wave energy may be required to provide satisfactory conditions inside the enclosed area. Most major breakwaters have been built of rock fill.

7.7 Approach channels

The ship lock of a tidal power scheme would have to be located so that ships approaching or leaving the lock would not be embarrassed by the discharges from the turbines or sluices. Unless the bathymetry of the seabed in the estuary is unusually favourable, the lock (or locks) would have to be located outside the main deep water channel, this channel being the normal route for ships but now occupied by the turbines. In order to provide access to the locks from either side at all states of the tide, approach channels may have to be dredged. These channels would normally have a width about seven times that of the largest ship, with additional widening at bends. If this involves dredging rock, then large costs will be involved. Again, extensive study and model tests would be needed to ensure a satisfactory and economical solution to the problem.

Chapter 8
Closure of a barrage

8.1 Introduction

Before the start of the 1978–81 studies of the Severn barrage, there was concern as to whether the completion of the final stages of construction of a Severn barrage would be feasible because of the currents to be expected in the final gap in a barrage across an estuary with a large tidal range. Consequently, the UK Department of Energy appointed the Netherlands firm of consulting engineers, NEDECO, to carry out a brief study of this aspect in the light of their experience in the Delta flood prevention works, where a number of tidal estuaries had been closed with barriers designed to prevent a repetition of the disastrous flooding which occurred as a result of the exceptionally high tide in January 1953 (Ref. 1978(1)).

In their report, NEDECO concluded that closure was feasible but would involve concrete blocks weighing 50 t at the final stages, this being the size needed to withstand the currents through the final gaps. Concrete blocks were proposed because rock of this size is virtually impossible to obtain. In the report of the Severn Barrage Committee, the maximum size of rock required for the final stages of embankment construction is about 2·5 t, a size which is relatively easy to obtain and to transport. What is the explanation of this fundamental difference in opinion?

In essence, NEDECO were asked the wrong question, if only because the complete closure of the Severn estuary by a barrage would not be acceptable on environmental grounds—the change from a strong tidal regime to a static one where the water was then progressively changed to fresh water by river flows and rainfall would have dire consequences for aquatic life in particular. The correct question would have been based on the concept of a barrage with a large number of openings available through which water would flow to and fro under controlled conditions. Thus the barrage would not have to be closed as such, simply built in such a way that the tides would be progressively diverted through the openings for the turbines and sluices until the barrage was complete and power generation could commence. This concept had been proposed in a brief study of a tidal power site in north-west Australia (Ref. 1976(1)) and was indeed used for the closure of one of the estuaries in the Delta works, namely the Volkerak in 1969, based on caissons with temporary openings (Ref. 1970(1)). Nevertheless, it is convenient to refer to the closure of the final gap in a barrage as 'closure'. This remains the most critical stage in the process of gaining control over the tides.

8.2 Calculation of water flows

One of the first reports produced for the Severn Barrage Committee addressed the problem of closure in detail (Ref. 1979(3)), using a mathematical model of water flows similar to those which have been developed for the design of river closures for the construction of dams to create reservoirs. Mathematical models of tidal flows are discussed in more detail in Chapter 9. For the closure model, the sea outside the barrage was represented by a simple sinusoidal curve of the same amplitude as the existing tide at the proposed barrage site, and the basin was represented by a depth/volume curve deduced from bathymetric charts. This type of model calculates flows through the part-complete barrage from the difference in water levels each side, and then calculates the change in basin water level from the flow, the length of the time step considered and the surface area of the basin at the levels in question. The part–complete barrage is represented by a series of equations relating flow to the difference in water level across the barrage. A series of equations is necessary so that each of the elements of the barrage which is open to water flow is represented correctly. For a single opening, it is a simple matter to calculate the flow for an assumed difference in water level across the barrage. This will be of the form

$$Q = C_d A (2gh)^{0.5}$$

where C_d is the coefficient of discharge, A the area of the opening and h the difference in water level across the opening. For accurate results, care has to be taken in selecting C_d and h. For the turbine water passages, which are asymmetrical in the flood and ebb directions, and assuming that the turbines are not in place, the following values of C_d were assumed in Ref. 1979(3):

$$C_d(\text{ebb}) = 1.94$$
$$C_d(\text{flood}) = 0.97$$

If the turbines have been installed, then flows are defined by the runaway characteristics of the turbine in each direction of flow. The turbine control and lubrication arrangements have to be designed to accommodate this method of operation. If this is not possible at reasonable cost, then the turbines that have been installed at the time of closure have to be suitably restricted in operation.

The design of venturi sluice developed for the Severn barrage and described in Chapter 5 includes a temporary water passageway above the normal passage, this providing additional discharge area to help reduce differences in level across the barrage during the later stages of closure. This feature would not be necessary with radial gates, where the whole of the water depth above cill level is available for flow, albeit at lower efficiency of discharge.

For the main water passageway of the venturi sluice, Table 8.1 gives typical values of the discharge coefficient C_d appropriate to use in a 'flat-estuary' model:

For the upper waterway, a discharge coefficient of 1.0 is appropriate, with the available head being reduced by the relatively small losses due to friction along the length of the channel. These losses can be calculated using one of the classical channel friction formulae.

An important feature of tidal power schemes is that, if the scheme is a large one, there will be many sluice and turbine openings which will be located in

Table 8.1 Values of discharge coefficients of venturi sluice

	Ebb	Flood
Classic design	?	1·65 (throat area) 0·91 (exit area)
Severn design	0·97 (throat)	1·8 (throat) 0·87 (exit)

groups to suit the geometry of the estuary. Consequently, it will not be valid to consider the flow through an opening in isolation; the flow through each group should be considered. Thus the water approaching and then leaving a group will form a wide jet which will be much slower to dissipate than the jet from a single opening, because the water available for mixing with the jet will be limited to that on each side of the jet.

As a result, the correct calculation of the flow through an opening must taken into account the kinetic energy of the water approaching and then leaving that opening.

Fig. 8.1 illustrates the difference between total energy and the water surface on each side of a sluice. The principal loss of head occurs in the length downstream of the throat, where the gently expanding 'draft tube' is designed to recover as much as possible of the large velocity head developed at the throat that would otherwise be lost through turbulence. The exit velocity will still be relatively large and so, in a relatively shallow estuary where the opening occupies a large proportion of the water depth, further recovery of kinetic energy occurs beyond the outlet.

Upstream of the opening, the approach velocity will have the effect of reducing the head required to accelerate the water into the opening. Thus, the flow through the opening will be given by

$$Q = C_d A_t [2g(H_u - H_d)]^{1/2}$$

where H_u is the total energy of the approach flow, and

$$H_d = h_2 + \frac{Q^2}{2gA_2 2} [(A_2/A_e)^2 - 2A_2/A_e + 2]$$

In passing, the turbine operating head is given by $H_u - H_d$.

For gaps in the barrage, which may comprise sections where no construction has started and sections where embankments are part complete and form cills over which the water flows, in both cases at varying depth during the tidal cycle, flows will be either super-critical or sub-critical. The former are governed by the water level upstream relative to the invert or bottom of the water passage, the latter by the net head across the barrage after deducting friction losses.

Thus, for critical flow, where the water depth at the inlet to the channel is less than 67% of the upstream total energy level minus friction losses in the channel, the flow is given by

$$Q = C_d A [0.67 \, g (H_u - h_f)]^{0.5}$$

where h_f is the friction head loss and is given by

$$h_f = n^2 L V^2 / (A/P)^{4/3} \text{ and}$$

V = velocity at the point where the water depth is 67% of the upstream total energy level

A = area of the gap

P = perimeter of the wetted area of the gap

n = Manning's friction coefficient, typically around 0.035 for a rough channel

L = length of the channel along which the water flows

H_u = upstream total energy level

g = acceleration due to gravity.

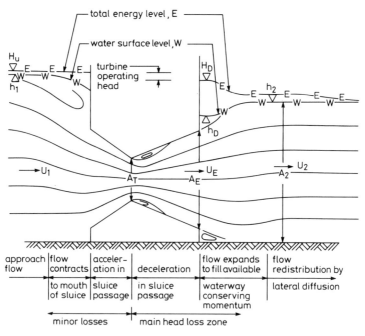

Fig. 8.1 Water levels and total energy across an orifice

For sluice, $Q = A_T \times \text{coefficient} \sqrt{2g(H_U - H_D)}$

Where H_U = total energy level of approach flow

$$H_D \approx h_2 \cdot \frac{Q^2}{2gA_2^2} \left[\left(\frac{A_2}{A_E} \right)^2 - 2 \frac{A_2}{A_E} \cdot 2 \right]$$

For sub-critical flow,

$$Q = C_d A [2g(H_u - h_f - H_d)]^{0.5}$$

where H_d is the downstream energy level.

Returning now to the computer model, the total discharge during a time step, estimated for each type of available opening from the head difference across the barrage, is used to calculate the change in water level in the basin. By iteration, the mean flow during the time interval, the change in tide level and the change in basin water level are balanced and the calculation moves onto the next time step. With a number of openings of different hydraulic performance, or whose hydraulic performance changes as the water level changes, perhaps as an opening dries out, the calculation becomes complex.

8.3 Construction sequence

Assuming that the principal elements of a barrage, namely the structures housing the turbines and those forming the sluices, would be floated into position, then the largest and most valuable elements should be placed first, before the natural tidal currents have been increased significantly by the partial blockage of the estuary caused by the barrage. Thus the turbine caissons should be placed first, followed by the sluice caissons and then by any plain caissons and, finally, by the embankments. This straightforward logic is complicated if the barrage is to be sited across a commercial shipping route, because one or more ship locks will be required and these will have to be operable to provide a safe route past, or through, the part-complete barrage as soon as the barrage begins to encroach on the normal shipping channel. Since ships normally navigate along the deepest water in an estuary, it follows that the turbine caissons, which occupy the deepest water, could begin to obstruct shipping soon after construction has started, and therefore work will probably have to start with the ship lock(s). On this basis, elements will appear in the estuary in the following order:

- Temporary protection for the ship lock
- Turbine caissons without turbines
- Sluice caissons without gates and with maximum area open to tidal flows
- Plain caissons, including those forming transitions between the end caissons of blocks of turbine or sluice caissons and embankments
- Embankments on erodible areas of the seabed
- Remaining embankments

The above order can be expected to vary with circumstances at a particular site. For example, there may be a need to protect an area of the seabed or foreshore from the erosion that would occur if it were left exposed to tidal flows of increasing strength until the end of construction, in which case either the appropriate length of embankment would have to be built early, or appropriate anti-scour protection provided. If a barrage is to be located at a site where there are extensive deposits of erodible sediments across the estuary, then it may be necessary to protect the whole width of the estuary against scour before any other construction starts. The width (in the direction of flow) of this protection need not be unduly large as long as the protection is of the 'falling apron' type which will deform to accommodate scour at its edge and thus prevent further scour underneath itself.

If a barrage contains a large number of turbines, say more than 20, then the production of these could take longer than the production of the turbine caissons. In addition, the cash flows will be adversely affected if all the turbines are built and paid for but not being used until the barrage is complete. Thus, as a general rule, only something over half the turbines should be ready to operate at the time that 'closure' takes place. Some of the remainder will be being assembled or connected up; the remainder will be being manufactured and so their water passageways will be available to contribute to the total area of the barrage open to tidal flows. These factors will have to be taken into account in calculating the flows through the various elements of the barrage.

8.4 Specific results

The information that is essential to those planning the various stages of construction of the barrage will comprise:

- The time available at high water or low water during which currents are less than the maximum that the equipment handling float-in elements can handle safely
- The maximum difference in water level either side of the barrage immediately after each element has been placed, and during the following few tides before ballasting and foundation sealing are complete
- The maximum currents through remaining gaps
- The changes to the normal levels of high and low water, both at the barrage site and elsewhere in the estuary.

Clearly, this information will be specific to the site being considered. In order to illustrate the effects of progressive construction of a large barrage on tidal levels and current speeds, results are taken from Ref. 1980(5) which refined earlier work done using a 'flat-estuary' model (Ref. 1979(3)). At the site considered, the spring tide range is 11·2 m and the neap range 5·8 m. The barrage tested had 140 9 m diameter turbines and 160 sluices of modified venturi type with vertical-lift wheel gates 12 m square. The sequence of construction follows that discussed above, starting with the temporary works for the ship locks and following with the turbine caissons and the sluice caissons, as shown in Fig. 8.2. Results are summarised in Table 8.2 and illustrated in Figs. 8.3–8.5.

Figs. 8.3 and 8.4 show that the effects of 'closing' a barrage with a large number of open water passages on tide levels is slight, even at advanced stages of construction.

Fig. 8.5 compares the maximum difference in water levels either side of the barrage with the percentage of the cross section of the estuary at the barrage site that is blocked. Different results are obtained for ebb and flood tides because the sluice and available turbine water passages assumed have different flow characteristics in each direction. The sluice passages, in particular, were

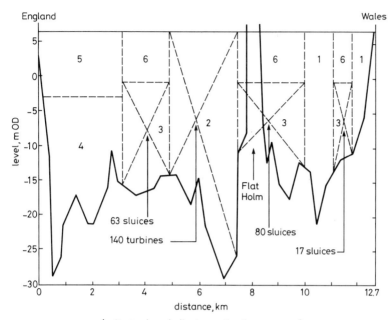

Fig. 8.2 Closure sequence as tested for the Severn barrage

designed for optimum performance when refilling the basin, and therefore contribute to the increased difference in water levels on the ebb. What is apparent is the severity of conditions during spring tides compared with neaps, pointing to difficult placing operations being confined to neap tides.

Fig. 8.6 shows how the maximum velocities at the barrage increase as the estuary is progressively blocked. Again, conditions during spring tides are far more severe than during neaps. In Table 8.2, these velocities are translated into the times during which velocities are less than 0·5 m/s or 1·0 m/s. The lower limit has been suggested as a 'safe' working window for the final floating into place and ballasting down of turbine and sluice caissons; the latter would represent a more confident but more risky approach. To set these in context, the drag force on a caisson caused by a current of a certain strength may be estimated by the formula:

$$F = 1{\cdot}5 \ A \ dH + 1{\cdot}5 \ A \ V^2/2g$$

where A = submerged area of caisson obstructing the flow

dH = difference in water level across the barrage

V = velocity of current

A factor of 1·5 is used to allow for the local increases in currents around the previously place caisson and to provide some factor of safety. In addition, the force should be multiplied by the relative density of sea water. This can be assumed to be 1·03.

Table 8.2 Hydraulic conditions at the Severn barrage during closure

Closure stage	0	1	2	3	4	5	6	7	8	9
Area blocked										
Flood %	0	11	14	33	24	54	62	66	66	65
Ebb %	0	11	15	26	30	62	70	77	79	78
Max. head difference (m)										
Spring flood	0	0.05	—	0.20	0.19	0.72	1.08	1.35	1.50	1.42
Spring ebb	0	0.04	—	0.14	0.17	0.75	1.19	1.56	1.77	1.67
Neap flood	0	—	0.02	0.05	0.05	—	—	—	—	0.40
Neap ebb	0	—	0.02	0.04	0.04	—	—	—	—	0.55
Max. embankment gap velocity (m/s)										
Spring flood	1.63	2.11	—	2.87	2.75	4.36	4.49	4.91	5.73[x]	5.04 (5.1)
Spring ebb	1.47	1.90	—	2.38	2.59	4.23	4.32	4.88	5.47[x]	5.44 (5.2)
Neap flood	0.81	—	1.10	1.52	1.42	—	—	—	—	2.67

Neap ebb	0.77	—	1.05	1.31	1.39	—	—	—	—	3.13
Time (min) gap velocity <0·5 m/s										
HW spring	55	42	—	33	27	18	17	13	6^{x+}	12
LW spring	48	36	—	24	27	66$^+$	240$^+$	234$^+$	276^{x+}	12
HW neap	150	—	99	75	71	—	—	—	—	30
LW neap	144	—	102	75	75	—	—	—	—	30
Time (min) gap velocity <1 m/s										
HW spring	120	87	—	66	60	37	35	25	17^{x+}	22
LW spring	138	87	—	60	63	84$^+$	250$^+$	244$^+$	282^{x+}	26
HW neap	00	—	258	120	162	—	—	—	—	60
LW neap	00	—	297	189	189	—	—	—	—	66

— No data
+ Cill dries at low water
x Condition in concrete passage above sluice
* Conditions in turbine caisson gap recorded with embankment closure conditions bracketted
00 Velocity does not reach 1 m/s

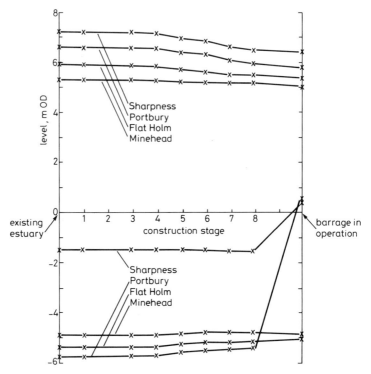

Fig. 8.3   Effect of closure on spring tide high and low water levels

8.5 Closure using rockfill

In most estuaries that could be developed for tidal power, there are shallow areas near each shore where it would not be practical to float in plain caissons. These parts of a barrage would logically be built using embankments where the first stage would be the building of an initial bank of rock to form a control structure, as discussed in Chapter 6. Once this was above the level of spring tides, the relatively still conditions behind would allow the remainder of the embankment to be built using dredged sand protected from wave attack on the basin side by mounds of smaller rock, minewaste or other low cost material. An important aspect of the control structure will be the definition of the rock size required at any stage of its construction.

Each load of rock, whether placed from a barge or from land-based plant, should have a size range or grading which will be stable against the currents and waves expected before the next load is placed or, in the case of rock on the seaward side of the control structure, until the main wave armouring is placed. Wave sizes will relate to the exposure of the site and the pattern of winds. Uncertainty and therefore risks will be reduced if the placing of the main wave armouring is kept close behind the leading edge of the advancing control structure. This will require plant capable of working in most of the weather conditions to be expected at the site.

The size of rock needed to withstand a particular current velocity has been the subject to much study in connection with the closure of rivers at the beginning of the building of dams. The most widely accepted formula relating velocity to stone size is (Ref. 1979(3)):

$$V = k [2 \ g \ d \ (s-1)]^{0.5}$$

where V = mean velocity through the gap (m/s)

k = a constant

d = stone size (m)

s = relative density of stone (noting that sea water has a relative density of about 1·03)

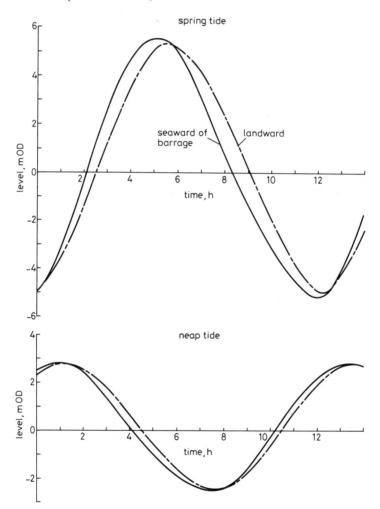

Fig. 8.4 Water level differences across the barrage at closure (stage 9)
—— water level d/s of barrage
--- water level u/s of barrage

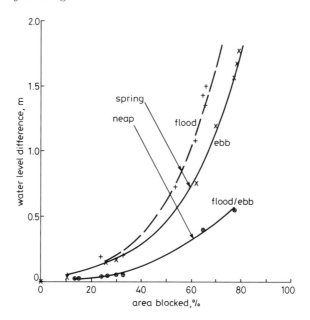

Fig. 8.5   Maximum water level differences during closure

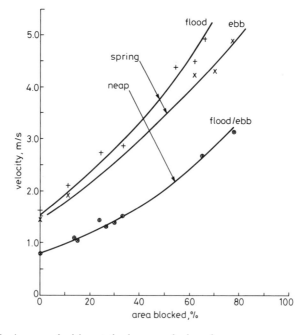

Fig. 8.6   Maximum velocities at the barrage during closure

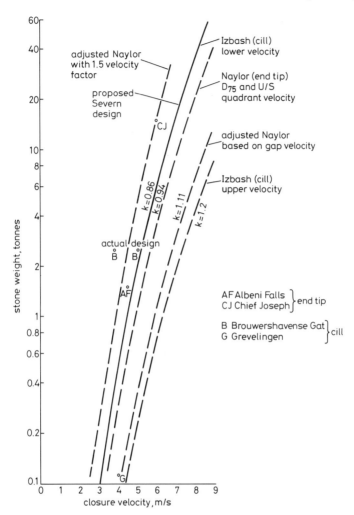

Fig. 8.7 Relationships between stone weight and closure velocity

An upper bound for k is about 1·2 where the stone is embedded in a level bed of rock fill, and the lower bound is 0·86 for a stone which is exposed. These results and other more recent results which generally corroborate these values are discussed more fully in Ref. 1979(3). A summary is shown in Fig. 8.7, together with the relationship recommended for design in Ref. 1979(3). These demonstrate the great sensitivity of the rock size needed to achieve control over tidal flows to the maximum velocity to be expected; doubling the velocity from 3 to 6 m/s increases the rock size from 0·1 to 6 t, a factor of 60. This explains the drop in stone size required to 'close' the Severn barrage achieved by allowing the barrage to have a large number of openings open to flow at the time the last embankment is being built, as discussed at the beginning of this chapter.

Chapter 9

Mathematical models of water movement

9.1 Introduction

The investigation of tidal power cannot proceed very far before some form of mathematical model, using a computer, of water flows into and out of the basin is required. The aspects which can be studied and evaluated in suitable mathematical models include:

- The energy output of the barrage
- The effects of the barrage on water levels in the basin
- The effects of the barrage on water levels to seaward
- Comparison of different methods of operating the barrage
- The effects of the part-built barrage on water levels and tidal currents
- The effects of the barrage on water levels during river floods and exceptionally high tides
- The effects of the barrage on tidal currents across and along the estuary
- The effects of the barrage on sediment movements
- The effects of the barrage on water quality
- The effects of the barrage on waves.

The above list is not exhaustive; almost any aspect of a tidal power barrage could be addressed by means of a suitable model. This chapter concentrates on models of water movement, which are fundamental to the understanding of tidal power.

There are four basic types of mathematical model of water movements, namely:

- 'Flat-estuary' models or O-D models
- One dimensional (1-D) models
- Two-dimensional (2-D) models
- Three dimensional (3-D) models.

As can be imagined, these are listed in order of increasing complexity and cost, and some judgment is needed when selecting the types of model to use during the study and design of a barrage. Each type is described and discussed below.

9.2 Flat-estuary non-optimising models

This type of model is the simplest and cheapest to develop and operate. The model is based on the assumption that a volume of water let into the basin will raise the level of the basin by an amount equal to the volume let in divided by the area of the basin at the time, i.e. spread uniformly over the basin. Thus the basin is defined by a simple area/depth curve, which can be calculated from navigation charts or other suitable source.

At the same time, the tide on the seaward side is assumed to be unaltered by the flows into or out of the barrage, although coarse corrections can be made, based on the results from more sophisticated models.

Having defined the basin behind the barrage as an area/depth curve, and the tide curve on the seaward side of the barrage, all that remains is to define the performance of the turbines and sluices which control the flows into and out of the basin. In each case, this is done by relating flow to head. In the case of the sluices, this can be in the form

$$Q = A C_d (2gh)^{0.5}$$

In the case of the turbines, their performance when acting as sluices to refill the basin during the flood tide can be defined in the same way as for sluices. When they are acting as turbines, their performance will be defined as two or more equations relating head to flow, efficiency and hence power output. If, as is normal, the power output at high heads becomes limited by the capacity of the generator, then the flow cannot increase further as the head increases and, instead, has to be reduced to stay within the capacity limit. This can only be assessed with the aid of turbine performance characteristics from the manufacturers with relevant expertise. Fig. 3.13 shows typical performance curves for a turbine selected for a particular scheme. These curves have to be converted into equations relating flow, power and head. The simplest method is to use a curve-fitting routine to identify the best cubic equation for each curve. The equation, and the limits over which that equation applies, can then be included in the model.

Having defined the performance of the turbines and sluices in the form of simple equations, it is reasonably straightforward to assemble a model which has the tide on the seaward side of the barrage starting at high water of, say, a mean spring tide and continuing through to a mean neap tide. After high tide, when the sea is ebbing (for an ebb generation barrage), the starting time of the turbines can be varied, with the resulting energy output calculated up to the time when the next flood tide has reduced the head across the turbine to the minimum at which they can operate. The starting time, and head across the turbines, which results in the maximum energy output will be the optimum for that tide. The process is then repeated for each tide in succession.

Fig. 2.2 shows a typical plot of water levels on the seaward and basin sides of a barrage, starting at mean spring tides and working through to mean neap tides. The instantaneous power output when generating is also shown, typically 5–6 hours during spring tides, about 3 hours on neaps.

This type of model can also be used to check approximately other methods of operation, for example:

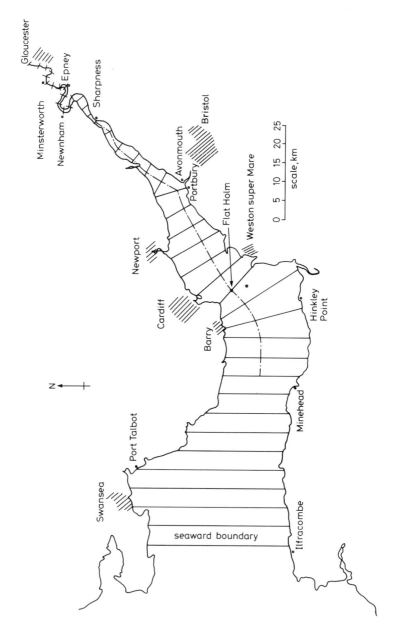

Fig. 9.1 Model section locations: Severn barrage ——— assumed estuary centreline

- Using the turbines to pump from the sea into the basin during a short period after high tide, when the difference in level between the ebbing sea and the basin is relatively small.

- Opening the sluices before levels have equalised, thus letting more water out of the basin (and reducing power output), which may be desirable for environmental reasons.

- Operation to maximise the value of the energy produced, particularly when this value varies during the day and when there is value in achieving an element of firm power at certain times of the day.

- Multiple basin schemes, examples of which are described in Chapter 12.

This type of model is quick to operate on a desk-top computer. Although important simplifying assumptions are built in, a large number of tests can be carried out quickly and cheaply. This type of model is therefore very useful at the start of a study, enabling alternative barrage arrangements, different numbers and sizes of turbines, their generator capacity, different types of turbines, different operating principles and difference sluice areas to be tested and compared. In addition, it is simple to include values of electricity which vary during the day in the optimisation process.

The extent to which the results from this type of model will differ from the real world (or from results from more sophisticated models) depends largely on the size and location of the barrage. If the barrage is located at the entrance to a small or short bay or estuary, then the tide on the seaward side should not alter much, and there should not be large dynamic effects along the length of the estuary. If, on the other hand, the barrage is located well up a long estuary, the tide on the seaward side could well be alterd substantially by the barrage, and there could be important dynamic effects in the enclosed basin. The Severn barrage is a case in point; the reduction in tidal range to seaward of the barrage, combined with the fact that the heights of high water and low water vary over the length of the estuary, result in predictions of energy outputs using flat-estuary models being about 20% optimistic. It is the absolute values which must be treated with caution.

9.3 Flat-estuary optimising models

The flat-estuary models described above can be used to optimise one or two aspects of operation during a tidal cycle; for example the starting time or starting head for the turbines. By covering a range of alternative options, other optima can be identified, such as the number of turbines or generator capacity, with reasonable confidence depending on the importance of dynamic effects in the estuary.

The next logical step is to avoid having to define the method of operation of the turbines as a single relationship between head, flow and power. This requires knowledge within the model of the full available range of operation of the turbines in the form of a 'hill chart' relating head, flow and efficiency, including their characteristics as pumps if reverse pumping is being considered.

Before the advent of digital computers, mathematical techniques for identify-ing the optimum route through the turbine hill chart during a tidal cycle were developed by Gibrat (Ref. 1966(1)). These have been incorporated into computer programs by various organisations working in the field of tidal power, particularly at the University of Salford where much of the early work on energy studies for the Severn barrage was carried out (Ref. 1979(7)).

In order to identify a maximum mathematically, an equation is needed which can be differentiated. Setting the differential equal to zero and solving the resulting new equation will find the maximum and minimum values. Gibrat's method therefore relies on the turbine hill chart being represented by a polynomial of the form

$$Q = ax^3 + bx^2y + cxy^2 + dxy^3$$

For a complex hill chart, the accuracy of a single polynomial will be variable. Consequently, when seeking to identify the optimum operating path, there will be inaccuracies. The method is not suitable when taking into account variations in energy value.

Consequently, to identify the optimum method of operation and take into account external factors such as variations in energy value, environmental constraints on water levels, turbine 'ramping' to avoid over-rapid rates of rise in power output, and so forth, so-called 'dynamic optimising' models have been developed.

This type of model requires a reasonably powerful computer with a large memory, simply because a very large number of operating options are evaluated and compared in order to identify the optimum.

Fig. 9.2 illustrates the method. Firstly, the basin volume is divided into layers of equal volume, the top of each layer being defined as a 'state' of the basin. This volume is selected as the minimum volume that the turbines can reasonably pass in one time step, say three minutes. Thus the turbines at maximum discharge at that head could empty the basin by perhaps six layers. It follows that the volume of a layer will be relatively small, equivalent to a change in basin water level of perhaps 10 mm.

In operation, the computer is programmed to evaluate all options available at the start of each step. During the generating phase these will normally include:

• No flow through the turbines

• A number of different flows through the turbines, each of which will change the water level in the basin by a whole number of complete states, and will be at a different efficiency.

In order to be able to consider different flows at each head, the complete turbine hill chart must be available, including limitations due to cavitation and generator rating. A method used by Binnie & Partners for studies carried out for the Severn Tidal Power Group is based on the non-dimensional hill chart being digitised and at the same time converted into a fine grid of co-ordinates relating head, flow, turbine efficiency and generator/transformer losses, and hence power. Submergence cavitation limits are included so that, if the turbine flow being tested causes the cavitation limit at that head and downstream submergence to be exceeded, that flow is not available.

In operation, this model considers all available operating options for moving from the start of a time step to the end of that time step. Thus at the end of the

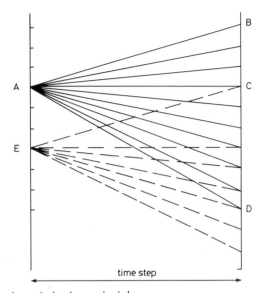

Fig. 9.2   Dynamic optimisation: principles

A = one of possible (or acceptable) water levels in basin at start of time step

AC = turbines shut

AB = turbines pumping maximum water into basin (also EC)

AD = turbines generating at maximum flow

If operation via EC results in more net energy output (or value) than route via AC, EC is preferable and AC is discarded. Thus optimum route is established for each state between B and D and calculation moves onto next time step.

time step, the basin water level could have been unchanged or lowered by between one and say six states or, if reverse pumping is an available option, raised by one or more states. This process has to be repeated for that time step for each state available from the preceding time step. Thus, if the water level in the basin can be changed during one time step by say six states, after two time steps 36 options will have been considered. Clearly, if an optimum method of operation is to be identified for even a single tide, covering perhaps 250 time steps, a very large number of options will be available and an efficient sorting routine is needed if the amount of data to be stored is to be manageable. This becomes even more important if optimisation is to be achieved over a full sequence of 29 tides.

The first stage in sorting out the options comprises identifying the operating path to each water state which results in the maximum energy output, or value, depending on which is to be optimised. If reverse pumping is being considered, then the path to each state above normal basin level is identified which requires the minimum energy (or value) input. All other paths to that particular state must be sub-optimum. The next stage looks at the end of the next sluicing period to identify the overall path which results in the maximum energy output or value for each final basin level, when the sea and basin levels are equal. In practice, the range of levels that can be reached at the end of sluicing is narrow, even if the level in the basin at the end of the previous generating period has been varied over a wide range as a result of the large number of options considered. If a single tide is being considered, the optimum path for each starting state which results in the basin level finishing at the same state is identified. It is much better to optimise over a complete spring–neap cycle because this takes into account the fact that succeeding high water levels are not normally equal. The optimum path when the next high water level is higher will not be the same as that when the next high water level is lower. At the end of the optimisation, the overall path can be identified which results in the maximum energy output or value, as appropriate.

If extra value is given to power being generated between certain times of the day, when electricity demand reaches a peak, this can be incorporated into the optimisation process by ascribing additional value to units generated between those times. In this case, normal operation will be distorted somewhat. As a general rule for a barrage operated as an ebb generation scheme, because the energy available in the water stored behind the barrage is lost if that water is not released through turbines before the next tide, the amount by which normal operation is distorted in order to achieve optimum overall benefit is slight. If reverse pumping from sea to basin is available, then a high capacity credit will result in normal operation being much distorted, depending on the relationship between the time of high water of spring tides (and hence mean and neap tides) and the times during which the capacity credit is available.

The computing effort and time required can be reduced by restricting carefully the range of water levels over which the optimising routine can search, for example, by putting an upper limit on basin level consistent with flood defence levels.

As an indication of the number of calculations to be carried out by this type of optimising model, a neap–spring–neap cycle will take about 5 hours CPU time on a 4 MIPS (million instructions per second) computer.

Because this type of model does not include the dynamic effects of the operation of the barrage on water levels in the estuary, the results must be interpreted carefully. Particularly important will be the effects of the barrage on the shape of the tide curve to seaward, especially at around low water when the turbines will be discharging large flows into water that is relatively shallow and still. At the barrage, starting the turbines, whether as turbines or pumps, will cause changes in water levels which will adversely effect the head across the turbines, and hence energy output or input. The first problem can be reduced by using a modified tide curve as predicted by a suitable dynamic model, but this will not take into account the variations arising from alternative operating paths being tested in the optimising model. The second problem has to be addressed in a dynamic model.

9.4 1-D non-optimising dynamic models

This type of model represents an estuary as a series of slices and combines these with the principal gravity, inertia and friction forces governing the movement of water to reproduce the progress of the tide into and out of the estuary. Ref.

1980(4) describes in some detail the development and proving of a 1-D model of the Severn estuary which has been used extensively to simulate barrages at various locations. The principal features of this type of model are as follows:

The model is 'depth-integrated'; i.e. the assumption is made that water velocities do not vary with depth. In a high-energy tidal regime, this assumption is a reasonable approximation. The model is also 'width integrated', so that velocities across the estuary at any cross section are assumed uniform. This assumption will be reasonable for an estuary which is narrow in relation to its width, but will not be appropriate in a wide embayment nor at the barrage where the flows through the sluices and turbines will be separated across the estuary.

The seaward limit of the model should be set sufficiently far away from any proposed barrage site that the effect of introducing the barrage on the tide curve will be minimal. In the case of the Severn barrage this distance was about 65 km (Fig. 9.1), where the Bristol Channel widens into the Irish Sea. For smaller estuaries, this limit could be near the mouth of the estuary, depending on the location of the barrage within the estuary.

The width of slice is related to the time step used in the model by the Courant stability criterion such that:

$$Dt > \frac{Dx}{(gH_{max})0\cdot5}$$

Thus, if the width of slice is 5000 m and the maximum depth is 30 m, the time step should not exceed 291 s or 5 min.

The flow is assumed to be one-dimensional, i.e. any flows not parallel with the axis of the estuary are ignored. The importance of this assumption will depend on the shape of the estuary. In a long, narrowing estuary like the Severn estuary the cross currents are not important in terms of energy, but, at the barrage, the asymmetry of flow caused by the different flows through the turbines and sluices will effect the head across the various water passages and will be important when considering sediment movements, navigation through barrage locks, and other local effects. Where a barrage is proposed at the mouth of an estuary, the sea cannot be represented satisfactorily by a 1-D model.

As well as currents flowing parallel with the axis of the estuary, this type of model ignores any vertical variation in water flows. Again, this is not important when considering energy output, especially as the flows near the barrage will be well mixed, but will be important when considering stratification or suspended sediments in the upper estuary.

The performance of the turbines and sluices is represented as single relationships between head, flow and power, in the same way as described earlier for flat estuary models, with the important exception that the transfer of momentum across the barrage is taken into account. Thus, for the sluices, the flow Q at head H will be given by

$$Q = AC_d(2gH)^{0\cdot5}$$

and H will be defined by

$$H = Y_u + (Q_t/A_u)^2/2\,g - Y_d - [(Q/A_2)/2\,g]\,[(A_2/A_e)^2 - 2A_2/A_e + 2]$$

where

> A = area of throat of sluice
> A_u = flow cross section area immediately upstream of the barrage
> A_2 = depth just downstream of a sluice multiplied by the total width of the sluice structure
> A_e = exit area of draft tube
> C_d = coefficient of discharge
> g = gravity
> H = total head upstream minus total head at exit from draft tube
> Q = discharge through one sluice
> Q_t = total discharge through the barrage
> Y_u = water level upstream of the barrage
> Y_d = water level downstream of the sluice after the initial expansion of flow to fill the full depth of water available

The equations representing the continuity of mass and the conservation of momentum for gradually varying one-dimensional flow can be written as follows.

Continuity of mass: $dA_s/dl + dQ/dx = q$

Conservation of momentum:

$$\frac{dQ}{dt} + \frac{d}{dx}\frac{Q^2}{A} + gA\left(\frac{dh}{dx} + S_f\right) = \frac{Q}{A}\frac{dQ}{dx}\left(1 - \frac{B}{B_s}\right)$$

where

> A = cross sectional area of flow
> A_s = storage cross section area
> B = surface width of flow cross section
> B_s = surface width of storage cross section
> q = lateral inflow per unit length
> $S_f = \dfrac{Q^2}{A^2}\dfrac{n^2}{R^{4/3}}$ i.e. the friction slope (Manning)
> n = Manning's coefficient
> R = hydraulic radius of the cross section (area divided by wetted perimeter)
> l = time
> x = distance

The equations have to be solved by a suitable mathematical technique, the one used in Ref. 1980(4) being a four-point half-time-step-implicit method. This estimates the discharge and level at each section, using the governing equations, at the next half time step. The estimated values are then used to convert non-linear terms to linear expressions related to the values at the next full time step. The resulting matrix of equations, being linear, can be solved by a sparse Gaussian elimination technique.

Having assembled the model as a representation of the geometry of the estuary with a driving tide at the seaward boundary and, where appropriate, constant flows at the points where rivers enter the estuary at their tidal limits, the next stage involves the adjustment of the friction term until the model

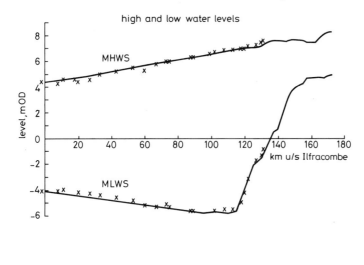

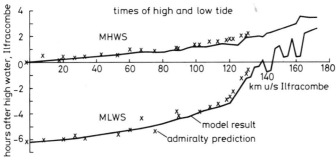

Fig. 9.3 Severn estuary: 1-D model calibration results with a predicted mean spring tide

represents the ebb and flow of the tides satisfactorily. Initially, this can be done using tidal data from ports or other sources, and some knowledge of the bed of the estuary and thus its likely surface roughness. When the model is representing 'standard' tides satisfactorily, it should then be tested by running actual spring and neap tides at the model's seaward boundary and comparing the progress of the model tide with simultaneous measurements of the prototype tide.

The factors to be checked and compared are:

- The heights of high and low water along the estuary
- The times of high and low water along the estuary
- The changing shape of the tide curve along the estuary.

Figs. 9.3 and 9.4 are examples for the Severn estuary, taken from Ref. 1980(4). Fig. 9.3 compares the heights and times of high and low water along the estuary for a mean spring tide as predicted by tide tables. Although the heights are represented very well, high water arrives slightly too early in the upper part of

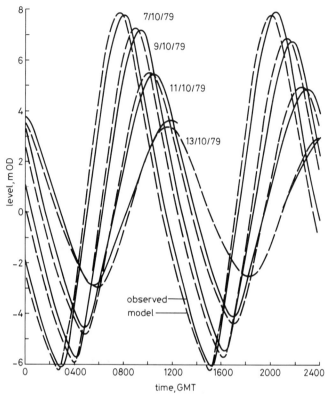

Fig. 9.4 Comparison of observed and model tides for October 1979 at Portbury
———— observed
— — — model

the estuary. Fig. 9.4 shows a series of different tides at Portbury, comparing model tides with actual measurements. The height and shape of the tide are represented well, but the time of high water is slightly earlier than it should be.

With a 1-D model developed and verified, reasonable confidence can be placed in the results of tests with a barrage in place. Figs. 2.5 and 9.5 show typical operating cycles for a spring and a neap tide. The immediately obvious differences when compared with results from a flat estuary model are the rapid changes in water level as the turbines are stopped and started. Other differences are the change in the shape of the tide to seaward of the barrage, and the effect that water ebbing from the upper parts of the basin has in raising high water level behind the barrage after the sluices have shut. Each of these features will affect the energy output of the barrage as estimated by a flat estuary model. Fig. 9.6 summarises the relationship between tidal range at the barrage and at the model's seaward boundary and energy output. This is virtually a straight line.

Having assessed the effects of a barrage on the dynamics of water movements in the estuary, a 1-D model can be used to provide a first indication of the effect of the barrage on the movement of non-cohesive sediments. This can be done by

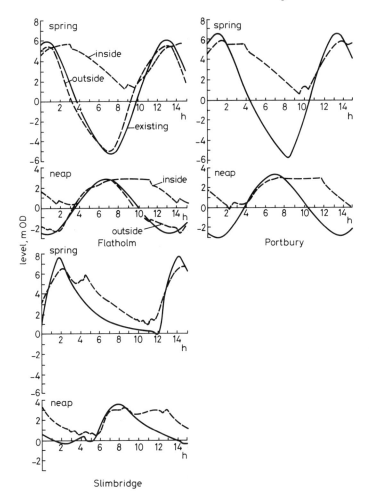

Fig. 9.5 Tide curves before and after barrage
—— existing
—— with scheme

including a sediment-transport function in the model, with the sediment size being set at the size found at each section of the model. This raises the need for appropriate field data on which to base the model, including:

- The size distribution of sediments in the estuary
- The rates of transport of sediments in the estuary.

Obtaining data on the sizes of sediments along an estuary need not be difficult or very expensive; simple grab samples of the seabed, taken at known locations and analysed in a laboratory, will suffice at this stage. Reliable data on the rates

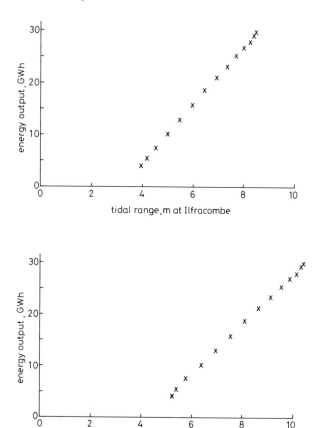

Fig. 9.6   Effect of tidal range on energy output for ebb generation
a tidal range at Ilfracombe
b tidal range at barrage

of sediment transport are much more difficult to obtain, bearing in mind the need for the data to be simultaneous, i.e. to cover the whole estuary during the same tide. In practice, quasi-simultaneous data will be much easier to obtain, i.e. data for tides of similar range. Storms and large river flows can also distort normal patterns of sediment movement. The equipment needed to measure rates of transport is sophisticated and expensive, very much the field of specialist laboratories such as Hydraulics Research Ltd. at Wallingford.

As an example of the difficulty of obtaining adequate data on sediment movements and matching these with a dynamic model, Fig. 9.7, taken from Ref. 1988(9), compares measured and predicted rates of sediment transport at a possible site of a tidal barrage on the estuary of the river Loughor near Llanelli in south Wales. Here the flood tide rises much faster than the tide ebbs, so that the rate at which sediment, in this case fine sand, is transported during the flood

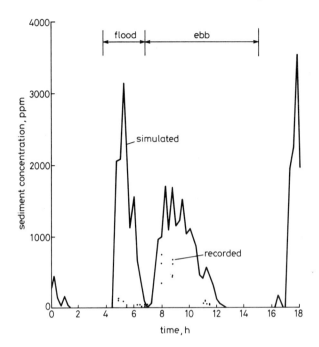

Fig. 9.7   Accuracy of sediment transport simulations (Longhor Bridge, spring cycle)

is predicted by a 1-D model to be much greater than during the ebb. As can be seen, this result differs from measurements taken at the site. The logical conclusion is that the site measurements were not representative of actual conditions, perhaps because the strength of the flood tide was enough to raise the sediments into suspension over the full depth of water.

The predictions discussed above all relate to potential sediment movements, i.e. indicate the rate of transport if enough sediment is available at that location. If the seabed happens to be bare rock, there will not be any sediment available. Similarly, if the actual sediment is finer than assumed, transport rates will be higher than predicted by the model. Furthermore if, in practice, large quantities of sediment are eroded at a particular location, this would lower the seabed, increase the depth of water and reduce velocities until a new equilibrium was reached (or a layer of non-erodible material was exposed). To predict this type of change with a computer requires a much more sophisticated model than that being discussed here. However, because the risk of sedimentation of the basin is a key aspect in assessing the feasibility of a tidal power scheme, it is useful to be able to predict broad changes relatively early and at low cost, based on a 1-D dynamic model which includes a sediment transport function.

Having defined the overall post-barrage hydraulic regime, some aspects of water quality can be addressed using a 1-D model. These include salinity, dissolved oxygen and biochemical oxygen demand (BOD), which are important for biological activity.

9.5 1-D optimising dynamic models

The flat-estuary optimising model described earlier cannot take into account the dynamic effects of different methods of operating the turbines and sluices on water levels. The 1-D model described above relies on a single turbine operating path; i.e. at any given head, there is a single turbine flow, and efficiency, available. In a new development for the Severn Tidal Power Group, a 1-D optimising model was developed and used to identify methods of operating the turbines and sluices which resulted in the maximum energy output and, separately, the maximum energy value from a Severn barrage.

In order to reduce the amount of computing time needed, the 1-D model reported in Ref. 1980(4) was simplified to 15 slices instead of 45. This still represented the dynamics of the main estuary with good accuracy, but the model was truncated above Sharpness because the water volumes above are insignificant in terms of energy. The optimising routine from the flat-estuary model was incorporated, with the main modification that the step between 'states' was a volume of water (6×10^6 m^3) equivalent to the layer 12 mm thick on which the flat estuary model was based.

This model was used to test ebb generation, ebb generation with reverse pumping at high tide from the sea to the basin, and early reverse pumping, before water levels had equalised. The results have not yet been published. What emerged was that turbine operation would become much more complex in order to take advantage of, or create, relatively rapid variations in water levels on either side of the barrage. Whether the turbines would be operated in this way, requiring frequent adjustments of blade angles, remains to be seen.

This type of model is complex, with a program length of about 6000 lines and requiring about 10 hours CPU time on a 4 MIPS computer to complete a full neap/spring/neap optimisation. One difficulty when studying the results is that there is no indication as to how different, or better, the identified optimum method of operation is compared with the second best method. The model selects a mode of operation which, hopefully, results in greater net energy output or value than that predicted by a similar 1-D model based on a single turbine operating path, but there could be a number of other paths, each of which may be attractive for some practical reason, which would result in almost the same energy or value.

9.6 2-D dynamic models

In a 2-D depth-integrated model, the estuary is divided into a grid of squares with the seabed represented as an average level along each side. Water can flow across any side of a square and its movement is calculated using the same basic techniques described above for 1-D dynamic models, but with the added sophistication that eddy viscosity and Coriolis force can be included. Turbines and sluices are also represented in the same way. Various organisations have developed and applied 2-D models to estuaries where tidal power schemes are or have been considered; e.g. Refs. 1979(5), 1981(4), 1988(7). The former Institute of Oceanographic Sciences (IOS), now the Proudman Oceanographic

Laboratory (POL), in particular has developed 2-D models covering very large areas of the sea (Refs. 1980(12), 1981(18)) and has also developed a 3-D model of the Severn estuary. Ref. 1980(3) also gives a good summary of the development and proving of a large 2-D model of the Severn estuary. Fig. 9.8 shows part of a plan of the Humber estuary as included in a 2-D model (Ref. 1989(2)). Because the grid size has to be small if the estuary is to be represented with reasonable accuracy, this type of model has to work on a short time step, typically about 1 min.

A 2-D model offers the following advantages over a 1-D model:

● Water movements that are not parallel with the axis of the estuary can be modelled. These can be important for navigation near the barrage.
● The different effect of the flows through the turbines and the sluices can be investigated.
● Coriolis force, which becomes more important as the area being modelled increase, can be included.
● Residual currents can be investigated; these are important when considering sediment movements.

On the other hand, there are complications when applying a 2-D model to an estuary with a large tidal range. Firstly, much more data are needed to prove and verify the model's representation of the strengths and directions of currents. Secondly, it is difficult to model satisfactorily the drying out of intertidal areas as the tide ebbs, because the depths of flow across grid squares becomes very shallow and the velocities too fast. This effect can be seen in Fig. 9.8. Thirdly, there may be features of the seabed which are important in terms of tidal flows but are narrower than the grid cell size being used. Low water channels running across the intertidal foreshore are an example. Similarly, rivers or narrow deep channels entering the estuary cannot be represented satisfactorily using a 2-D grid, so that these have to be added as 1-D sub-models.

The first problem, the availability of enough suitable field data for proving and verifying the model, reduces to the availability of enough funding and the time needed to acquire the data. A comprehensive data set, covering at least a spring tide and a neap tide, including simultaneous measurements of the tide level and current strength and direction over the full depth of water at perhaps 20 locations, will require considerable resources. If sediment movements are to

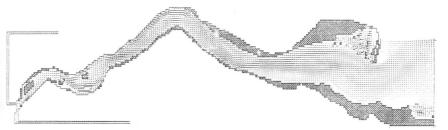

Fig. 9.8   Humber estuary: tidal current vectors predicted by 2-D model (Courtesy Humber Barrage Group)

be studied in the model, then appropriate simultaneous measurements will also be needed, from the seabed to the water surface, together with good data on the composition and properties of the seabed sentiments.

The second problem is perhaps the most severe: where there are large areas of drying sandbanks in an estuary, it can be very difficult to get a 2-D model to represent tide levels and currents satisfactorily. In 1989, studies were in progress at Hydraulics Research Ltd. to identify means of reducing this problem.

In Fig. 9.8, the existing conditions in the estuary near low water of an ebb tide are represented by arrows showing the strength and direction of the current in each grid square. The dark areas are dry sandbanks. Although currents are relatively slow in the mouth, the tide is still ebbing in the inner estuary.

Having established the changes that can be expected in the pattern of water movements, potential transport of non-cohesive sediments can be investigated in the same way as described above for 1-D models. At the time of writing, this has not been done in the UK—nor has any attempt been made to develop a 'moving bed' 2-D model which would allow the seabed level to adjust in response to the erosion and deposition of sediments.

The nearest that 2-D models have come to predicting sediment movements is by the extraction of bed shear stresses. As explained in Ref. 1988(8), mud is able to settle out of moving water if the bed shear stress is less than about 0.07 N/m^2 and will consolidate unless the bed shear stress rises to about 1 N/m^2 within a few hours. Similarly, recently deposited soft mud that has consolidated for a few days could be stable until the bed shear stress rises to about 3 N/m^2. By plotting contours of maximum bed shear stress during a spring and a neap tide cycle, again with and without a barrage, some indication can be obtained of the areas where mud can be expected to be eroded and the areas where it could accrete. These results need careful interpretation, taking into account the availability of mud for erosion and deposition, the effects of river floods in bringing sediments out of sub-estuaries into the main estuary, the effects of storms in eroding mud in shallow areas exposed to wave attack, and the effects of salt marshes, algae and other mechanisms for trapping or fixing suspended particles.

From this discussion, it will be clear that it is difficult to predict and analyse the effects of a tidal power scheme located in a large or relatively wide estuary without developing and proving a suitable 2-D model, but the collection of enough suitable data and the model itself will be much more expensive than for a 1-D model.

9.7 3-D models

One important weakness of the depth-integrated 1-D and 2-D models discussed in the previous Sections is that no account is taken of the variation of current strength with depth, nor of any vertical currents that may be present. Both these are important when trying to predict the effects of a barrage on sediment movements. The development and, more particularly, the proving of this type of model will be an order of magnitude more difficult than for a 2-D model.

The first 3-D model of a large tidal estuary, the Severn estuary, has been

Fig. 9.9   Physical model of The Wash (Hydraulics Research, Ltd.)

developed at POL and the model has been used to demonstrate the general effects of a Severn barrage. The results have not yet been published. In due course, a fine grid model of this type will be necessary to help predict the post-barrage fine sediment regime, which will be crucial to water quality and to the development of marine organisms forming the start of the food chain in the enclosed basin.

9.8 Physical models

This Chapter has concentrated on the application of computer models to tidal power schemes. The effects of marine projects on tidal flows have in the past, before the advent of computers, been investigated with the help of physical models. Fig. 9.9 shows the size and complexity of a model of the Wash, built by Hydraulics Research Ltd. on the east coast of the UK (Ref. 1974(4)). Physical models are still widely used, for example for port developments. Their use for tidal power projects has been limited; the most notable recent example being the use of an existing model for the preliminary assessments of the Mersey barrage (Ref. 1988(7)). One reason is that the barrage operation, particularly that of the turbines, is difficult to model accurately at the small scale needed. Another reason is that the time needed to set up and carry out a test is much greater, and therefore inherently more expensive, than with a computer model.

The future application of physical models to tidal barrages is likely to concentrate on aspects which cannot readily be modelled in a computer with adequate accuracy or confidence. One example is the placing of caissons at the barrage, for which an exceptionally fine-gridded computer model would be needed to represent satisfactorily the flow regime in three dimensions around and between caissons which are close together.

Chapter 10
Environmental aspects

10.1 Introduction

A tidal power scheme will reduce the tidal range in the basin it encloses and will therefore reduce the volume of water entering and leaving the basin during each tide, the tidal 'cubature'. These fundamental changes will have quite profound effects on the environment in its broadest sense. Predicting the changes in tidal range is relatively straightforward. Predicting the consequential changes in water quality and then on through the ecosystem, and then assessing their importance, is much more difficult. Separately, assessing the effects of building a barrage and its subsequent operation on the human environment is largely a qualitative exercise. In this chapter, an attempt is made to describe the various changes and set them in perspective.

10.2 Experience to date

There are only two tidal power schemes of any size operating in the world at the present time. These are the Rance barrage, near St. Malo in France, which has a capacity of 240 MW and started generating in 1966, and the 20 MW pilot scheme at Annapolis Royal in the Bay of Fundy, Nova Scotia, which was commissioned in 1985.

The Rance barrage was built at a time of uncertainty in France as regards future developments in power generation and was conceived as a large-scale prototype for a much larger scheme to seaward. No formal assessment was carried out of the possible environmental impact of building the barrage. In addition, the barrage was built in the dry, behind a cofferdam which almost completely blocked the estuary for several years. Consequently the salinity of the water in the basin dropped from that of sea water almost to that of fresh water, causing a fundamental change in the types of aquatic life in the basin. This change was then reversed when the barrage was completed and the cofferdam removed. Thus the experience gained at La Rance is only partly relevant to tidal power schemes elsewhere. More recently, studies have been made of the types and numbers of flora and fauna and so forth, the ecosystem, in the basin.

At Annapolis Royal, a single large turbine was installed in a concrete power house which was built in an existing island. The island is linked to the banks of the estuary by a causeway which includes sluice gates. These already existed and were built to regulate the water level in the enclosed basin for the benefit of farmers. Thus conditions in the basin were already somewhat artificial before the tidal power scheme was added. However, studies are being made of the changes in the ecosystem arising from the new hydrodynamic regime.

10.3 Water quality

There are many factors which affect the quality of the water in a tidal estuary, some natural and some due to human activities. Table 10.1 summarises some of the more important ones together with the predicted effects of a tidal power scheme. Each has been the subject of much study in connection with the Severn barrage and an extensive bibliography is available. Some of the main points are discussed below.

10.3.1 Salinity

Seawater contains about 32 grams per litre of sodium chloride or common salt, plus traces of a large number of other minerals. In a tidal estuary, the salinity varies as a result of fresh water inflows from rivers and, to a small extent, rainfall. This variation covers the full range from 'undiluted' seawater outside the estuary to fresh water at the tidal limits of rivers and tributaries. Fig. 10.1 shows this variation for the Severn estuary together with the effects of changes in river flows, and the predicted effects of the Severn barrage (Ref. 1980(2)). This highlights two fundamental points:

● Over a length of about 60 km, nearly half the length of the estuary, salinity varies naturally between about 30 g/l and about 10 g/l

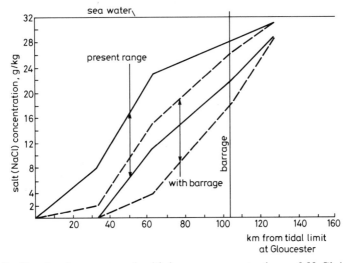

Fig. 10.1 Simulated present and with-barrage concentrations of NaCl in Severn Estuary

Table 10.1 Some criteria concerning water quality

Criterion	Main factors influencing criterion	Likely effects of a barrage	Possible remedial measures	Relative importance
Salinity	Tidal cubature	Cubature reduced by 20–50%		
	Freshwater flows	Boundaries between sea water and fresh water move seawards inside basin		
		Aquatic life readjusts		
Turbidity	Tidal currents	Away from turbines and sluices currents will be reduced significantly so turbidity will reduce	Beneficial for aquatic life, so no measures needed	High; influences wide range of aspects
	Waves	Some reflection outside may affect shores near barrage on seaward side. In basin, waves concentrated at upper part of shore	Use non-reflecting construction, embankments instead of plain caissons	
	Supplies of fine sediments	may erode muddy areas	Protect muddy areas with anti-scour systems	

Table 10.1 *Continued*

Criterion	Main factors influencing criterion	Likely effects of a barrage	Possible remedial meaures	Relative importance
Biodegradable pollutants, e.g. ammonia	Tidal cubature	Slight, because reduction in cubature partly offset by greater dilution at low water due to raised minimum water levels	Reduce or treat sources of pollutants	Moderate in polluted estuaries Low in clean estuaries
	Turbidity Density of human population and quality of sewage treatment	Reduced turbidity allows sunlight to penetrate water, promoting growth of algae which feed on wastes	Improve sewage treatment/disposal	Beneficial to aquatic life, as long as algal blooms are infrequent
Conservative pollutants, e.g. metallic wastes	Tidal cubature Freshwater flows	Up to 100% increase during droughts	Reduce discharges of wastes	High in estuaries supporting heavy industry
	Adsorption by sediments	Reduced sediment mobility could reduce amounts of pollutants trapped		

Table 10.1 *Continued*

Criterion	Main factors influencing criterion	Likely effects of a barrage	Possible remedial measures	Relative importance
Dissolved oxygen	Tidal cubature/ dispersion	Slight changes in main estuary; depending on resuspension of anaerobic sediments during spring tides	Improve sewage treatment	High; important for aquatic life, especially fish
	Water temperature			
Bacteria	Sewage inflows			Moderate for water sports and beaches
	Penetration of sunlight	Reduction in dispersion largely offset by greater dilution at low water. Reduced turbidity will be beneficial in allowing sunlight to kill bacteria	Improve sewage treatment	

● With an ebb-generation barrage, a similar variation will occur over a similar length of the estuary but the length affected will move further seaward about 20 km.

The predictions were based on an assumed reduction in 'dispersion coefficient' of 60%, this being a measure of the ability of the tidal waters to mix and disperse dissolved matter. For example, a small dispersion coefficient could result in the water in the estuary stratifying, with salt water entering on the flood tide sliding under the fresh water because of its greater density. The true value for the dispersion coefficient will vary from point to point around the estuary and will depend on tidal range and the pattern of water velocities and water depths. A reduction of 75% was considered to be an upper bound in order to obtain an estimate of the maximum likely changes in salinity and other indications of water quality.

10.3.2 Turbidity

Turbidity is a measure of the amount of particulate matter that is in suspension in water. These particles, which can be either of organic or inorganic origin, are kept in suspension by a range of mechanisms including turbulence due to high water velocities, convection, and changes in density of the water. In well mixed tidal waters the most important causes of turbidity are high water velocities due to both the ebb and flow of the tides, and waves in shallow water.

For water to be turbid, there has to be a source of fine particulate matter. This obvious statement opens up a range of relevant aspects. For example, why are some estuaries relatively free of muddy deposits and thus have relatively clear water? La Rance is an example of a relatively clean estuary in spite of having a tidal range similar to that of the Severn estuary. This question is addressed later. Extensive measurements of suspended sediment concentrations in the Severn estuary, carried out by IOS and others (Refs. 1981(11), 1982(2)), have been analysed to show that the amount of fine, largely inorganic, sediment in suspension during a high spring tide in the winter could be 30 Mt, dropping to about 3 Mt during a low neap tide. Where does this sediment come from, and what would be the effect of a barrage on these figures?

The first part of the question is fairly simple to answer. The amount of fresh sediment entering the estuary has been estimated at about 1 Mt a year, most of this being supplied by the rivers. Although this is an approximate figure, what is clear is that it is minute in comparison with the amounts in suspension, which are material which is travelling up and down within the estuary with the tidal flows.

The second part of the question is much more difficult to answer for two reasons. Firstly, there is little relevant information available from similar projects. Secondly, the modern solution, which would be to develop and apply a mathematical model of turbidity to the problem, has not, at the time of writing, been applied to a tidal power scheme. Consequently, reliance has to be placed on judgment, which is dangerous when there is little practical experience of this particular problem.

In simple terms, a tidal power scheme halves the tidal range in the enclosed basin. Thus the first assumption that could be made is that the turbidity in the basin would be about the same as that under present conditions with a tide of

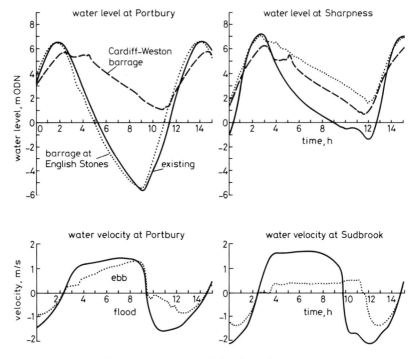

Fig. 10.2   Effects of Severn barrage on tide levels and currents

half the range in question; i.e. the future turbidity during spring tides would be similar to that during present neap tides. This suggests a reduction by a factor of about 10 during spring tides. Beyond this, one can only speculate until a model of turbidity is developed, accurately validated, and applied to a proposed tidal power scheme. This speculation could be along the following lines.

The ability of a flow of water to lift sediment into suspension is highly dependent on the velocity of the water. In a shallow estuary, i.e. one where the tide curve is no longer sinusoidal but has a steeper rise during the flood than the fall during the ebb, the flood tide has the strongest currents and thus will be dominant in mobilising fine sediment into suspension. Because a tidal power barrage is designed to be transparent to flows during the flood, in order to fill the basin as high as possible, the rate of rise of the flood tide in the basin will be only slightly less than in the undeveloped estuary. Consequently, the amount of fine sediment in suspension should not decrease as much as suggested by simple comparison of tidal range.

At present, after a short period of quiet conditions at high water (ignoring waves for the present), the ebb tide generates currents which are less strong than those during the flood (Fig. 10.2 from Ref. 1987(14)). With a working barrage, there will be a longer quiet period at high water, and the strength of the subsequent currents in the basin generally will be much less than at present. Consequently, there will be much more opportunity for sediments to settle and consolidate at high water, and the power of the ebb currents to remobilise them will be greatly reduced. This points to a mechanism where fine sediments in deep water will be pushed up into shallow water during flood tides, then will settle out of suspension and not return on the next ebb. With the amount of new material being relatively small, a tentative conclusion is that the inorganic turbidity in the basin will decrease steadily as the material available in deep water is deposited in the shallows around the water's edge and does not return. The principal mechanism available for re-suspending foreshore mud will be waves, another, less important, being heavy rain. It is known that prolonged wave attack from one direction can erode large quantities of sediments. Thus prevailing wind directions will be important in causing a second distribution of fine sediments, moving them from exposed shores to sheltered areas.

So far in this thesis, only hydrodynamic factors have been considered. There are important natural forces available which will assist the process of removing sediment from suspension and thus lowering turbidity in the basin. Firstly, algae have been shown to have an important role in 'fixing' mud on the intertidal foreshore. Secondly, saltmarsh vegetation traps suspended sediment which has been carried into it during spring tides, by slowing down the rate at which water escapes, during the ebb tide, between its stems and roots. On the other hand, if inorganic turbidity is reduced the penetration of sunlight will increase, allowing algae and phytoplankton to feed on dissolved nutrients. The algae and phytoplankton themselves cause turbidity. The development of algae is discussed in more detail later.

The tentative overall conclusion to be drawn concerning the effect of a tidal power scheme on the inorganic turbidity of the water in the basin is that the turbidity should slowly decrease to a level that can be sustained by the arrival of fresh supplies from the rivers feeding the estuary and from the sea, but will be liable to temporary increases during storms. In addition, the material lost from circulation will tend to end up on the intertidal foreshore in areas sheltered from waves generated by prevailing winds. The lowering of turbidity due to fine sediments will allow sunlight to penetrate much further into the water and so give algae and plankton the opportunity to develop.

Because of the importance of fine, cohesive sediment to the aquatic environment of tidal estuaries, much recent work has been done, or is in progress, to collect relevant data, establish the properties of fine sediments and to develop appropriate computer models. Examples are found in Refs. 1986(12), 1988(9), 1988(10) and 1988(12).

10.3.3 Biodegradable pollutants

Pollutants such as ammonia or bacteria tend to decay exponentially with time. Because of the work done around the world on improving the quality of tidal waters subject to pollution from untreated, or partially treated, wastes, the processes are quite well understood, if complex. Apart from simple dilution and dispersion near the point of discharge, these pollutants will be affected by various chemical and physical processes.

Mathematical models are widely used in forecasting the future behaviour of pollutants in estuaries. Their application to tidal power schemes is less well founded because of the lack of data against which the forecasts can be compared. Two important factors in the equation are the post-barrage turbidity, which will affect the growth of algae which feed on nutrients such as nitrates and phosphates, and dispersion. Both these have already been discussed. However, predictions made with a hydrodynamic model of the Severn estuary (Ref. 1980(2)) show that the overall effects of a tidal barrage would be small (Fig. 10.3) in comparison with the changes which occur already as a result of variations in tidal range and fresh water flow.

If the concentrations of ammonia, nitrates and so forth will be little changed by the introduction of a barrage, but the turbidity due to inorganic particles can be expected to be reduced progressively, then conditions would be much more favourable for the growth of algae and plankton. This has given rise to suggestions that the basin enclosed by the Severn barrage could suffer from 'blooms' of algae (Ref. 1988(4)). Certainly, sudden large increases in numbers of algae, called 'blooms', are a recurring problem during the spring and early summer in rivers and shallow reservoirs, causing problems for the water supply industry.

A complex mathematical model of the processes of natural growth and decay of the simpler estuarine organisms has been used to predict the effects of a barrage on the Severn estuary ecosystem. This is the GEMBASE (General Ecosystem Model) model developed by IMER (Ref. 1980(2)). In the absence of firm figures for the changes to be expected in turbidity and dispersion, dispersion was assumed to be reduced by 75%, the maximum reasonable estimate, and turbidity by a factor of about 25.

The model predicted that there would be large numbers of phytoplankton in the first few years after the barrage started operating. These would be food for carniverous zooplankton and both would be food for deposit feeders, such as worms, and suspension feeders such as mussels. Thus these higher organisms are predicted to increase in numbers; in the case of suspension feeders, which are almost absent from the inner estuary, by a factor of 100 although this development would be constrained to some extent by the seaward movement of the zone of low-salinity. This development would in turn benefit wading birds.

10.3.4 Conservative pollutants
Conservative pollutants are those which are not subject to biochemical decay. Salt is one example. The pollutants which are important in the context of a tidal power scheme are those arising from industries located around the shores of the estuary or along the banks of the rivers flowing into it. If there are no industries, then pollutants can nevertheless enter the tidal waters from the atmosphere but the importance of these is likely to be small.

One reason for heavy industry to be located on the shores of estuaries with large tidal ranges is that effluents are diluted and dispersed rapidly by the strong currents. Thus substantial quantities of metals such as cadmium, chromium, zinc and lead, and non-metals such as arsenic, are discharged to waste. However, the quantities have tended to decrease in recent years in the UK, partly due to the decline in heavy industry and partly to the use of more efficient manufacturing processes. In developing countries, industrial discharges are tending to increase in advance of measures to deal with the resulting pollution.

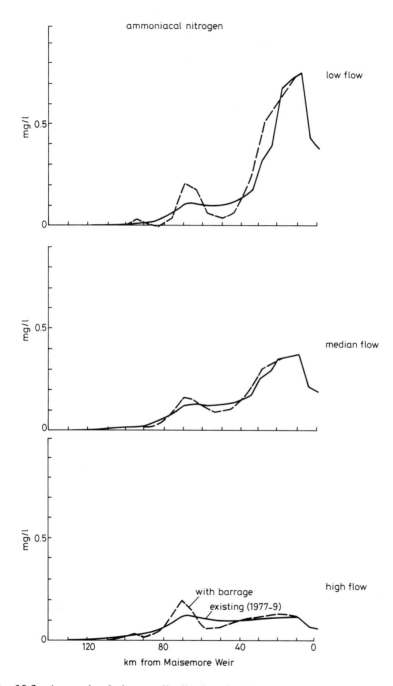

Fig. 10.3 Ammoniacal nitrogen distributions for three freshwater flows (Institute for Marine Environmental Research)

At first glance, the effect of a tidal barrage on a conservative pollutant will be to increase its concentration by the ratio of the amount of water entering and leaving the estuary before the barrage is built to that amount after the barrage is set to work, i.e. the inverse of the change in tidal cubature, or a factor of about 2. This ignores the fact that the tidal 'excursion', which is the distance covered by a particle during a flood tide or an ebb tide, although large, is a small proportion of the total length of a long estuary. For example, the total quantity of water entering and leaving the Severn estuary above the Holm islands during each tide averages about 4×10^9 m^3. The average fresh water flow in the rivers Severn, Wye, Avon, Taff and Usk totals about 30 m^3/s or $2 \cdot 5 \times 10^6$ m^3/day. Thus a particle released into the water will travel up and down the estuary for a large number of tides before reaching the open sea.

In the Severn estuary, the tidal excursion near the island of Flat Holm, which is close to the proposed site of the barrage, during spring tides is about 25 km and during neaps about 12 km. In the sub-estuaries such as the river Avon it is typically about 15 km during spring tides. The Severn estuary is about 170 km long from its tidal limit at Gloucester to Ilfracombe. The time for a particle released into the river at Gloucester to pass Ilfracombe will vary with the fresh water flow, but is about a year. Thus the main method of 'disposal' of industrial pollutants is dilution and dispersion rather than the simple flushing out with the ebb tide.

A further important factor is the location of the point of discharge of each pollutant along the estuary. In the Severn estuary, the main inputs of nickel arise from outside the estuary while the main input of cadmium is near Avonmouth. Because the Severn barrage would have little effect on the patterns of flow in the rivers near their tidal limits, the concentrations of nickel are predicted to increase by around 50% in the estuary during times of low or average river flows, and to change little during periods of high flows (Fig. 10.4). What this figure also demonstrates is the importance of river flows in diluting the nickel wastes in the first place.

For a pollutant such as cadmium which enters the estuary rather than the rivers, the river flows are less important. Fig. 10.5 shows the predicted effects of a barrage on cadmium concentrations for high and low river flows which are close to extreme values; the low flow conditions are those near the end of the 1976 drought. The maximum increase would be about 50%. What is not clear is whether this increase would be important when the 'normal' concentration is only around one part in 10^9. A much greater concentration is allowed in drinking water. It is known that shellfish seem to be able to extract pollutants such as cadmium from the water and deposit them in their shells. This removes the pollutants from circulation for some time.

The process of adsorption is complex. By mixing particles of undissolved pollutants with particles of fine silts and clays in suspension, electrostatic forces cause the pollutants to become attached to the silt and clay particles. When, in quiescent conditions, the silt and clay particles are able to settle out of suspension and consolidate, the pollutants are removed from circulation. This is an important mechanism in the apparent removal of pollutants from tidal waters. Ref. 1988(2) includes data on concentrations of metals in the sediments of the Mersey estuary. If one of the effects of a barrage is to reduce greatly the

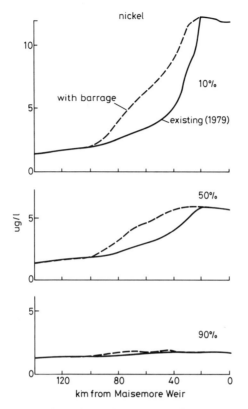

Fig. 10.4 Nickel distributions determined by IMER for three freshwater flow conditions (10, 50 and 90 percentiles)
––– with barrage ——— no barrage

amounts of suspended sediments, then the process of adsorption will be reduced and the pollutants will be in circulation longer, so that their concentrations can be expected to increase.

10.3.5 Dissolved oxygen

The amount of oxygen dissolved in the waters of a tidal estuary is crucial to marine life living in, or passing through, the estuary. In the latter category fall migrating fish, particularly salmon and sea trout, and eels, all of which are important commercially.

Dissolved oxygen concentration is one of the aspects of water quality with which there is considerable experience in mathematical modelling. For the Severn barrage, model predictions have been made by IMER after first comparing the concentrations in the estuary at present as predicted by the model with measured values. Fig. 10.6 summarises the results and shows that the effects of a barrage would be negligible, except for a slight reduction from normal values over a length of about 30 km of the upper estuary. It is

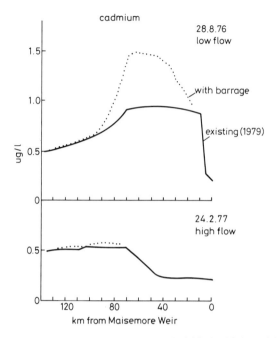

Fig. 10.5 Cadmium distributions determined by IMER for high and low conditions
—— no barrage ⋯ inner line

noteworthy that the amount of oxygen that can be dissolved in water decreases as the temperature of the water increases, so that concentrations in summer are lower than in winter.

10.3.6 Bacteria

One aspect of concern about tidal power schemes is the possible effects they may have on the dispersion of sewage effluents. In populated areas on the coast, sewage which may be partly treated or not treated at all is usually discharged into the estuary via a long pipeline or outfall. Old outfalls may discharge their untreated sewage a short distance offshore, with the result that the incoming tide can bring human faeces and other unpleasant visible wastes onto nearby beaches. More insidious will be the very large numbers of bacteria in the same water. These waters will degrade naturally, sunlight, for example, being important in destroying bacteria.

Fairly simple measures will prevent unsightly wastes being deposited on beaches, and appropriate steps are being taken in many countries where there are recreational beaches near cities. These include the relocation of outfalls much further offshore.

In the basin enclosed by a barrage, the raising of low water level to about present mean sea level is predicted to have a significant and beneficial effect on bacteria counts. Again, this prediction is based on computer simulations, which show bacteria counts at the shore near a long sea outfall which may serve

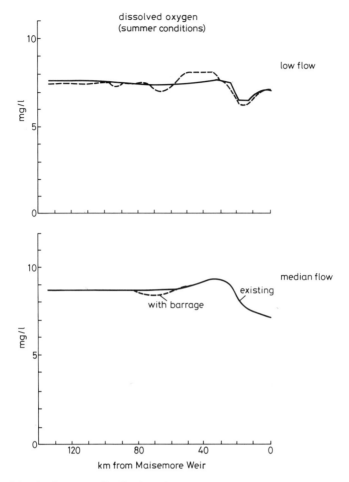

Fig. 10.6 Dissolved oxygen distributions for two freshwater flows with and without a barrage
——— no barrage ⋯ with inner line barrage — — — with outer line barrage

200 000 people dropping from about 100/ml to about 30/ml. For an old, short, outfall serving say 20 000 people, the effect of the barrage would be a small reduction (Ref. 1980(6)).

Overall the predictions that have been made by experts of the effects of a barrage on water quality are encouraging. Where there could be problems, these are primarily caused by other human activities such as chemical manufacturing industries which have been able to take advantage of the high tidal range to 'flush' waste products away. In the developed countries, a greater awareness of the results of such pollution coupled with the development of more efficient manufacturing processes has resulted in a tendency for the amount of pollution to decrease with time.

10.4 Sediments

Brief mention was made earlier in this chapter to the fact that the Rance estuary is relatively free of sediments in spite of having a tidal range similar to that of the Severn estuary. The fact that estuaries with large tidal ranges usually contain large quantities of sediments is demonstrated in the UK by the Severn estuary, Morecambe Bay, Solway Firth, the Humber and the Thames estuaries. Similarly, most small estuaries are so full of sediment that they virtually dry out at low water of large spring tides.

The story starts effectively during the last Ice Age. The increase in the size of the polar ice caps, which lead to the glaciation of large areas of land, caused the sea level to drop substantially, by up to 40 m. Thus river estuaries would have been much further to seaward than they are now. This can often be seen on navigation charts, where narrow deep water channels mark both the old course of the river and branches to tributaries.

With the end of the Ice Age sea levels rose again. The effects of shallow water in causing the incoming tide to steepen means that the flood tide has greater capacity to lift sediment into suspension than the slower ebb. In addition, sediment deposited near high water mark tends to be left behind by the mechanisms discussed earlier in connection with turbidity. Consequently, there was a mechanism for moving mobile sediments up estuaries. At the same time, the erosion of land, newly exposed as the ice retreated, by rain and wind supplied sediments to rivers which transported these down river until the speed of the water dropped enough to allow the sediment particles to fall out of suspension, if they were fine particles, or to stop rolling along the bed of the river if they were coarse particles. The widening and deepening of a river as it enters its estuary cause this drop in speed so that sediments accumulate until they have reduced the cross section of the estuary enough to prevent further deposition. The area of deposition would then move downstream. This simple mechanism is complicated by the variations in river flows; dry spells would allow sediments to be deposited in the river itself, while floods would move large quantities downstream.

Thus there were two mechanisms for moving sediments. One moved sediments upstream towards the top of the estuary, the other brought sediments down the river. The combined effect would be to fill progressively the estuary from the top towards the sea. The rate at which this took place would have depended on the quantities of sediments available, and this gives a clue why different estuaries with the same tidal range can contain very different amounts of sediments.

For some large estuaries on the coast of the UK, a major source of sediment has been shown to be offshore seabed deposits (Ref. 1987(11)), while the rate of supply of new sediments from the rivers has been estimated at only about 1 Mt/ year (Ref. 1982(2)). We are led to the conclusion that most of these sediments were moved up the estuary, as a result of the dominance of the flood tide over the ebb, as the sea level rose after the last Ice Age.

In the Severn estuary, there are relatively small amounts of sediments outside the Holm islands and Bridgwater Bay, although sheltered areas such as Swansea Bay are filled with sediment. This absence of sediment may be due to a lack of supply rather than to some inherent property of the estuary. This begins to give a clue as to why the Rance estuary is relatively clear of sediment in spite of its high tidal range. This is related to the overall shape of the English Channel, essentially a very large estuary, tapering down to the Straits of Dover, which in glacial times was the outlet of the river Rhine. Although the tidal flows through the Straits are far larger than would occur at the mouth of a river, the overall pattern of tides, with a small range at the 'entrance', at Land's End, increasing generally towards the 'top' of the estuary, is similar to that in normal estuaries. This large estuary has also been supplied with sediments from the west, in the same way as the Severn, and these have been deposited first in the uppermost reaches and in sheltered corners to seaward. This process has not yet reached the Rance estuary.

   This concept of estuaries being filled from their uppermost reaches downwards mainly by sediments brought in from seaward deposits can only be considered true in its wider sense, and therefore subject to various local variations. However, it does form a basis for assessing the possible impact of a tidal barrage on sediment movements. In addition, the movement of sand and silt particles can be represented satisfactorily in computer models of water movement, so that the changes caused by a barrage can be predicted with some confidence.

   Fig. 10.7 shows how large could be the changes in the rates of transport of sediment by tidal currents in the basin behind the Severn barrage (Ref. 1987(14)).

10.5 Effects on nature

So far, the predicted effects on a tidal power scheme on water quality and sediments has been discussed. These then affect the productivity of the tidal waters and consequently the higher forms of life including fish, birds and, of course, man.

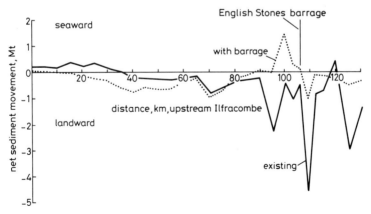

Fig. 10.7   Net potential and sand movement over a typical spring–neap–spring tidal sequence
——— natural estuary ··· English stones barrage

Before addressing these questions, it is worth pointing out that the direct environmental impact of a tidal power barrage is small; it does not cause acid rain or leave radio-active residues. There are no high pressures or fast-turning machinery, so the risks of explosions are minimal. There is very little noise involved when a barrage is working. La Rance barrage is a low structure about 740 m long. To produce the same energy in an area with reasonable winds would require about 100 very large windturbines sited at about 1000 m centres, i.e. an area approaching 100 km^2. The tides are, of course, predictable.

The food chain in a tidal estuary starts with dissolved and particulate nutrients, continues through algae and plankton, which have been discussed already, to filter feeders and deposit feeders such as molluscs and worms, then reaches the higher life forms such as fish, birds and mammals. Vegetation, particularly saltmarsh vegetation, forms an important sink and source of nutrients. We have already seen that the reduction in tidal currents and sediment movements caused by a tidal barrage should lead to large increases in the lower forms of life. The effects of these changes and the presence of a barrage on the higher forms are discussed next.

10.5.1 Fish

Assuming that the turbines of a barrage are not used in reverse to pump extra water into the basin at around high water, the main flow of the flood tide will be through the open sluices. These should not deter fish wishing to migrate upstream, such as sea trout, elvers and salmon, although open-topped sluices may be preferable to designs with closed-in tops. If the turbines are also used to refill the basin, by being allowed to freewheel in reverse, then fish will be drawn through these. Because the flows will be less than those during power generation, the stresses on the fish should be proportionately less. Thus a tidal barrage should not present undue problems for fish moving upstream through the barrage on the flood tide.

For fish moving downstream, there is only one sensible route, which is through the turbines during the time that power is being generated. The most important commercial species to be considered are adult eels and young salmon, called smolts. Before considering whether passage through the turbines would be possible, it is worth considering what happens to the water flow.

The speed at which the water in the basin approaches the entrance to the turbine water passages is slow, perhaps 1 m/s, depending on the depth of water upstream. This speed will be larger than the speed that can be sustained by small fish such as smolts, which is about 300 mm/s. Thus small fish will not be able to avoid being drawn into the turbine openings unless they can stay at the surface and end up in the relatively 'dead' areas immediately above intakes. If this happens, they will be at risk from predators, particularly gulls and, perhaps being unwilling to descend, may not continue their journey to the sea. Thus passage through the turbines could be the lesser of two evils. The overall shape of turbine water passages is designed to accelerate the water up to the turbine runner, where much of its energy is extracted, and then slow the water down as smoothly as possible so that its kinetic energy is converted back to potential energy. This maximises the effective head across the runner. Thus the approach velocity becomes about 2 m/s at the entrance. At the runner this has risen to about 10 m/s while at the same time the pressure has decreased to below atmospheric pressure, and then the velocity slows to about 2 m/s again and the pressure recovers. These figures are appropriate to a tidal range of say 7–10 m. Smaller changes would apply to smaller ranges. For a 9 m diameter runner, the total length of the water passage would be about 65 m and the time taken by a fish passing through would be about 20 s. In addition, at the 'distributor' or guide vane system upstream of the runner, the water flow is given a sudden twist so that it approaches the runner blades at the optimum angle. The runner blades act in much the same way as an aircraft wing, flying through the water and thus generating lift which turns the turbine shaft. When operating at best possible efficiency, the flow downstream of the runner is relatively smooth. When operating much below best efficiency, this flow is highly turbulent.

To summarise, a fish passing through the turbines would have to survive acceleration from 1 m/s to about 10 m/s and back, together with a temporary drop in pressure equivalent to a change in water depth of about 20 m. At the distributor, there are a number of radial vanes to be negotiated and the water is set spiralling. At the point of maximum velocity, the runner blades will be encountered edge-on. Thereafter, the water velocities and pressures return steadily to normal.

This apparently formidable obstacle course could be expected to wipe out all fish attempting it. In practice, experience gained on hydro-electric schemes on rivers has shown that fish are able to survive passage through smaller turbines, running at higher pressures and much higher speeds of rotation, remarkably well; i.e. survival rates of 85% of more (Ref. 1988(6)). Thus a tidal barrage should not present an undue obstacle. However, experience at the Annapolis Royal prototype turbine has shown that a proportion of the fish passing through the turbine are being beheaded. This may be being caused by the additional vanes downstream of the runner, which support the central shaft. These three vanes are located where there is considerable swirl in the water, especially when the turbine is operating at low efficiencies, and so fish may be colliding with these vanes sideways. Studies for the Severn Barrage Committee showed that it should be possible to design this type of turbine without this bearing, i.e. as a cantilever from the upstream support, and so omit the offending vanes.

If, as suggested earlier, smolts or other small fish stay on the surface and do not pass through the turbines, one possibility would be to provide them with fish passes, broadly similar to the fish ladders used to enable fish to circumvent dams on rivers. These would be more complex because the water level at each end, the sea and the basin, would be varying almost continuously and thus require some form of adjustable weirs to maintain a suitably attractive flow. Alternatively, it may be practical to collect the waiting smolts in suitable nets and simply lift them over the barrage.

If the turbines are to be used to pump water into the basin at around high water, then a considerable flow, and probably hydraulic noise, would be generated which could divert fish migrating upstream from the 'quiet' sluices. If this happens, then adult salmon in particular could be at risk from injury because of their size and because the turbine's runner blades will be upstream of the distributor vanes and thus not be so well aligned with the flow. The distributor vanes themselves will be in an hydraulically inefficient position.

There will be some risk that fish migrating upstream will be attracted by the noise of the turbines working normally during the generating period and try to swim upstream against an impossibly fast flow. Developments in sonic control systems may provide the solution by discouraging them from the area (Ref. 1988(3)). These and other possibilities are discussed by Solomon (Ref. 1988(6)).

Overall, although much further work and trials remain to be done, the conclusion can be drawn that a tidal barrage, with sympathetic design, should not present an unduly severe obstacle to migrating fish. If this cannot be shown to be the case, then owners of sport fishing rights upstream can be expected to object to any proposals to build a barrage.

Other types of commercial marine fish, such as plaice and flounders, should benefit from the lower turbidity and more stable sediments which should provide better conditions for breeding (Ref. 1980(18)).

10.5.2 Birds

Large tidal estuaries form a vital link in the migration patterns of a wide variety of wading birds. On their annual migrations, these birds fly vast distances. During their flights, reserves of fat are converted into energy, followed if necessary by the conversion of muscle, particularly leg muscle, until the next feeding area is reached. Thus wading birds arrive at estuaries where there are large areas of intertidal mudflats to feed on the invertebrates in the mud, and have in many cases developed specialised beaks to enable them to reach their prey. The numbers involved can be enormous, over 100 000 at a time in the best estuaries. Partly because they are at the top of the food chain, wading birds are seen as an indicator of the general health of an estuary. In addition, of course, they form a fascinating subject of study for many thousands of people and, with organisations such as the Royal Society for the Protection of Birds looking after their interests, the protection of birds is seen as an important duty of a civilised country (Ref. 1976(2)).

Having set the background, the effects of a tidal barrage on the bird life of the estuary can be considered. To seaward, there should be minimal change, because the tidal range will be reduced only marginally. This assumes that the barrage has been designed and located in a manner that does not result in wholesale changes to sediments outside the barrage.

Within the enclosed basin, two fundamental changes in the pattern of tides will occur, namely the raising of low water levels to about mean sea level, and a 'stand' at high water while the turbines await an adequate difference in water level between the sea and the basin to develop. The first change will result in approximately half the intertidal foreshore being permanently submerged and therefore no longer available as bird feeding area. This is the change that has caused most concern to those involved with the protection of birds because it implies that only half the food will be available, so that either birds will have to compete for the remaining food or they will avoid the estuary in preference for another.

The second change, the high water stand, will result in less feeding time being available, because the intertidal area will be exposed for about nine hours out of every 12·5 hour tidal cycle, instead of perhaps 12 hours, depending on the shape of the natural tide curve, and, conversely, the roosting birds will be close to the shore longer and therefore more likely to be disturbed. These worries have lead the RSPB to oppose the development of tidal power in any estuaries, and to publicise this view widely (e.g. Ref. 1987(5)). The picture may be less black than it appears, particularly in the estuaries with the largest tidal ranges, such as the Severn estuary, and following early studies (e.g. Ref. 1980(8)), much work is in progress in the UK and abroad to increase our understanding of the factors involved in birds' feeding patterns.

Considering now the Severn estuary, the area of foreshore that is exposed at low water of a high spring tide is vast, about 200 km^2. Yet the number of birds that feed in the Severn estuary is proportionately much less than in other estuaries with smaller tidal ranges such as Morecambe Bay or the Mersey estuary, in spite of the large intertidal areas available. One reason is that the tidal currents during spring tides are so strong that much of the mud is lifted into suspension. It has been estimated that up to 30×10^6 tonnes of sediment are mobile in the deep water channel alone (Ref. 1982(2)). This must disrupt the growth of the worms and molluscs which burrow in the mud, and studies have shown that those in the Severn are smaller and less populous than their cousins in other estuaries. After a barrage is built, the sediments will be much more stable, as discussed earlier, and the amounts of sediment in suspension will be greatly reduced. This can be expected to result in a large increase in the numbers and quality of deposit and filter feeders, to the general benefit of wading birds.

A further factor in the overall picture is the composition of the foreshore sediments. The sediments which are exposed near low tide are located in a stronger current regime than those towards the upper edges and thus tend to be sands rather than muds, and therefore less biologically productive. It is these sediments which will be permanently submerged rather than the 'useful' muds; so the effective loss of bird feeding area will not be proportional to the reduction of tidal range.

The change in the shape of the tidal curve at high water will reduce the time available for birds to feed and may leave them more at risk of disturbance from the shore. However, the post-barrage tide curve will be similar in shape to the natural curve at Southampton Water, where there is a pronounced high water stand due to two tides coinciding, one from the west the other the previous tide from the east. Southampton Water and adjacent estuaries such as Langstone harbour are hosts to very large numbers of wading birds. Thus a high water stand should not present undue problems. Another factor in the argument hinges on the fact that wading birds always feed at the water's edge. This is linked to the movement of the worms and molluscs downwards as the sediments are exposed. With a tidal barrage, the length of the shoreline will not change, so that birds will not be more crowded than they are at present. In addition, the slower ebb of the tide will mean that the birds will not have to walk so fast, and the reduced tidal range will mean that they will not have to walk so far while feeding.

Having discussed the mechanics of feeding, the base of the food chain that ends with birds should not be forgotten. To allow organisms to flourish, adequate supplies of food are essential. It is an unpalatable fact that some estuaries attract large numbers of birds because they receive large quantities of sewage effluents. The Mersey estuary is a prime example. Thus cleaning up an estuary, while of benefit to fish, can reduce its productivity and thus eventually lead to a reduction in bird numbers. For example, the cleaning of the waters of the Thames estuary, a major achievement in water engineering in the 1950s first allowed life to flourish in the mud, feeding on the vast amounts of organic debris available while being able to breath in the cleaner water. This bonanza attracted very large numbers of ducks, until a balance had been reached between the supply of food and the predators.

The overall conclusion to be drawn is that, in estuaries with very large tidal ranges, tidal barrages should not have a detrimental effect on wading bird populations and may well be beneficial.

A detailed assessment of the environmental effects of the Severn barrage was carried out by the Nature Conservancy Council during the studies carried out for the second Severn Barrage Committee (Ref. 1981(7)). The Council concluded that, on the evidence available at the time, it would oppose the building of the Severn barrage. This view is reasonable bearing in mind the duty of the Council to conserve nature. This duty means that, for example, any proposals to remove the half-tide cooling water pound built on the foreshore of the Severn estuary for Oldbury nuclear power station would be resisted. This pound is now a site of special scientific interest (SSSI) because it provides an important roost for birds.

10.5.3 Mankind

The effects of a tidal barrage on man's activities will be wide ranging, and quantifying these has in most cases to be subjective. For example, if you live in a house overlooking a tidal estuary, and enjoy the continuous changes in light and shade as the tide ebbs and floods, you would be entitled to object to this view changing but it would be difficult to put a value on the change. The market value of the house would be one measure, and this may increase because the estuary would be easier and safer to sail on, especially for small craft. Better grounds for objection could be the impact that large numbers of hard-living construction workers, imported to build the barrage, could have on your peaceful village. On the other hand, shopkeepers and landladies would probably welcome the increased revenue, especially after the barrage is completed and opportunities for water-based recreations are developed.

Starting at the beginning, the first effect of a tidal barrage on the estuary on which it is located will be the disturbance caused by its construction. Fortunately, there is a great deal of flexibility here, assuming that the site of the barrage has adequately deep water. The principal components of the barrage could be built elsewhere and floated into position, as discussed in earlier chapters. In addition, the embankments joining the main structure to the shore could be built almost entirely as a seaborne operation. This would result in the minimum noise and disturbance to towns or villages around the shores of the estuary. Alternatively, if the employment and general boost to the local economy offered by the construction project is welcome, then the barrage could be designed to be built largely within the estuary; turbines and other specialised equipment would normally have to be brought in from centres of heavy manufacturing industry.

The next effect is the visual intrusion caused by the barrage. Some intrusion is inescapable, because at low tide a barrage which is designed to have little clearance above high tide will be exposed by at least the tidal range. Those most affected will be those living close to the ends of the barrage and those using the shore or the sea close to the barrage. On the other hand, those who have visited the Rance barrage in France are generally not left with the impression that the barrage is more intrusive than, say, a low road bridge. Compared with other forms of generating electricity from renewable energy, tidal barrages can be considered unobtrusive, at least partly because they are compact in terms of the amount of electricity they generate.

A tidal barrage is quiet, so there should not be problems of noise.

Moving away from the barrage, the most obvious, indeed fundamental, effect as perceived by those living around the estuary or using it will be the change in the shape of the tide curve behind the barrage. The area of mud or sand exposed at low tide will be halved. Whether or not this is an improvement will be entirely a subjective matter, but, in general, people like the appearance of large areas of water and the resulting variations in light. As a result, for example, artificial lakes were formed to improve the views around many stately homes. The longer duration of high water is the next most important effect, and the same arguments apply.

Turning to less esoteric aspects, estuaries with large tidal ranges tend to be difficult places for boating and sailing. The currents are strong, which can make life difficult for small sailing craft, and many water-based activities are constrained by the need to start and/or finish near high water when there is adequate depth of water. Because an ebb generation barrage will raise minimum water levels to around present mid-tide levels, cause high water levels to be prolonged for two or three hours and reduce the strength of tidal currents in the basin generally, the enclosed basin will be much more accessible, and safer, for water-based recreation. Purists will point out that great skill is needed to sail on tidal waters, and that anybody can sail on an inland lake, so the 'improved' conditions behind a barrage will not necessarily be welcome. This view has to be respected. The successful development of Kielder Water, the largest man-made lake in the UK, for recreation was based on plans set out by Northumbrian Water Authority in Ref. 1978(2). Ref. 1980(11) was the first assessment of these benefits in relation to the Severn barrage. This report concluded that, although the opportunities for the development of recreation and tourism around and on the enclosed basin would be considerable, additional investment would be needed to provide the necessary facilities.

The effects of a tidal barrage on industrial concerns depends largely on the amount of shipping using the estuary, discussed in Chapter 7, and on those industries which rely on the tides to remove liquid wastes, discussed earlier in this chapter. Otherwise, the construction, operation and maintenance of a tidal barrage will bring work to appropriate industries.

Chapter 11
Economics of tidal power

11.1 Introduction

There are two aspects involved when considering the economics of tidal power. Firstly, the cost of the electricity produced and, secondly, the value of that electricity. If the cost is less than the value, then that tidal power scheme can be considered to be economic. However, other methods of generating electricity, if available, may be more economic and therefore preferable. These two aspects, cost and value, are discussed separately in the following Sections.

11.2 Costs

Everything being equal, the barrage at a particular location which will have the best chance of being built will be that which produces electricity at the lowest unit cost. Thus identifying the optimum barrage design, in terms of unit cost of electricity, is a task which is crucial and, unfortunately, somewhat complicated. The principal factors which have to be taken into account are listed on Table 11.1, together with their main interactions with other factors. The factors are listed in what is more or less the order in which they are considered if the process of identifying the optimum configuration of the barrage is to be carried out in a logical manner; they are not listed in order of importance as regards their effect on the cost of energy produced by the barrage.

The unit costs of construction materials should be established early, partly because this can be done without reference to what is happening elsewhere in the process, and partly because the results will influence much of the other work. To begin with, global costs will be the most useful; for example, the all-in cost of high quality reinforced concrete including formwork. This might be £500/m^3 in an industrialised country but perhaps £300/m^3 in a developing country because labour costs are low. Some of the principal materials are:

- Concrete and reinforcing steel
- Good quality rock (for embankments)
- Fabricated steel (for gates)
- Skilled and semi-skilled labour.

To begin with, while overall concepts and outline designs are being developed, broad comparisons, and thus judgments as to what are more likely concepts, will have to be based on these global costs. As the design develops, more detailed cost information, which takes time to collect, can be used in order to refine the preliminary thoughts.

Table 11.1   Factors involved in the optimisation of unit cost of electricity

Factor	Effects	Comments
Unit costs of construction materials	Overall, direct effect. Internally, affects choices of materials, e.g. use of plain caissons instead of embankments.	Initially, reliance has to be placed on all-in unit costs of concrete, rock etc.
Turbine type (bulb, Straflo etc)	Influences design and cost of power house.	Bulb turbines are more versatile and more proven. Straflo turbines are more compact. Tubular turbines may be suited to small schemes.
Turbine runner diameter	Influences design and cost of power house and associated electrical equipment.	Economies of scale apply, so that larger machines are cheaper per unit power, subject to enough water depth being available at low tide.
Method of turbine regulation	Energy production. Cost and complexity of turbines. Need for downstream flow-control gate. Capability of pumping at high tide, or reverse generation.	Double-regulated turbines are the most productive. Variable distributor allows downstream gate to be omitted. Two-way generation and/or pumping at high tide requires variable runner blades or double regulation.

Table 11.1   *Continued*

Factor	Effects	Comments
		Double regulation is more difficult with Straflo design.
Turbine rotation speed	Energy production. Submergence requirements.	Reduced speed reduces flow and power output, and also reduces required submergence.
Generator capacity	Energy production. Costs, including costs of transmission link(s).	Electrical losses at low powers are important.
Generator voltage	Transformer costs. Switchgear costs.	Second order aspect.
Number of turbines per electrical group	Costs of transformers and switchgear. Importance of load rejection.	Significant for large schemes only.
Number of turbines per caisson	Size of caisson. Placing methods and equipment, e.g. tugs.	Three turbines per caisson is reasonable minimum, to achieve stability when dewatering a water passage for inspection/maintenance.

Table 11.1 *Continued*

Factor	Effects	Comments
Number of turbines.	Energy production. Overall costs. Barrage layout.	Key part of optimisation. Optimum in terms of unit cost is reached when the addition of another turbine increases overall unit costs of electricity produced. Optimum economic scheme is reached when the unit cost of electricity produced by the next turbine exceeds that of the alternative source of electricity.
Type of sluice gate	Design of sluice gate structure. Hydraulic performance, and hence water levels in basin and energy production. Maintenance costs. Passage of fish.	For deep water (say 15 m+ at high water), vertical lift best. For shallow water, radial gate is viable alternative but less efficient.
Size/number of gates	Water levels in basin. Energy production. Overall costs. Ease of final 'closure' of barrage.	Key part of optimisation process. Maximising high water levels benefits ship movements as well as energy production. Increased sluice area reduces tidal currents in remaining gaps and thus simplifies 'closure'.

Table 11.1 *Continued*

Factor	Effects	Comments
Sizes/number of ship locks	Shipping travelling to ports in basin. Pleasure boat movements. Overall costs. Construction programme.	Locks represent the main 'non-productive' element. Locks have to be operational by the time that the construction of the rest of the barrage is beginning to affect safe ship passage.
Plain caissons and embankments	Overall costs. Construction programme. Wave reflection.	Plain caissons could be built in the same facilities as the turbine and sluice caissons. Embankments are cheaper option for shallow water. Embankments are less risky during final stages.
Cost of alternative forms of electricity generation	The easier the market, the greater the number of turbines that can be installed	Long-term projections of fuel costs for thermal power stations are subject to the greatest uncertainty. Tidal power development is a long-term development.
Interest rates	Economic viability of tidal power. Construction programme.	High interest rates favour low-capital/high-running-cost projects, and reduced construction programme.

The next aspect, turbine type, will have relatively little influence on overall costs, largely because turbines and their generators do not account for more than 30% of total costs, normally much less, and also because the small differences between different designs will not be important in terms of cost. Why this aspect has to be considered early is that much effort on caisson design and costing will depend on the decision concerning the type of turbine. Because most experience worldwide with tidal power and with low-head run-of-river hydro schemes has been gained with bulb turbines, this type is the logical first choice.

The choice of turbine runner diameter can and should be made early, again so that those designing the caissons have something to go at. If the estuary is deep, say 20 m or more at low water, the decision will be based on what could be considered the maximum proven size, because increasing size normally brings economies of scale. A runner diameter of about 9 m can be considered a reasonable upper limit. Otherwise, a runner diameter equal to half the water depth at the barrage site and low water of spring tides will be a reasonable starting point. If the estuary is shallow, consideration will have to be given early to the practicality of lowering the seabed at the barrage site by dredging. This opens up a range of associated questions concerning sedimentation which are better avoided if possible, at least at the beginning of an investigation. Thus estuaries or barrage sites which are unduly shallow should be avoided unless they are attractive for other reasons; for example a road crossing.

The method of regulating the turbines, i.e. controlling the water flow, affects both the energy output and the cost of the barrage. The most productive type, and most expensive, is double regulation. This is described in Chapter 3. If the barrage is to be designed to generate in each direction of flow, double regulation is essential. If the turbines are to do no more than generate in one direction of flow, single-regulated machines are feasible. Either the runner blades or the distributor vanes can be movable. The former type has a wider operating range which makes for smoother operation, but requires a downstream gate to stop and start the water flow. This has important implications for the designers of the power house.

At the start of a study, the assumption that the turbines will be double regulated will result in the maximum energy output being identified, with the final choice of method of regulation being left until a later, more detailed stage.

When considering the turbines and generators, some early judgements can be made which will have some influence on more important issues. These are the turbine rotation speed, the generator capacity, the generator voltage and the number of turbines in each electrical group.

For synchronous operation, the turbine rotation speed has to satisfy the equation:

$$RN/2 = F60$$

where R = speed of rotation (rpm)
N = number of poles on the generator rotor
F = frequency of supply (Hz)

Increasing the speed of rotation, which increases water flow and thus power output, reduces the number of poles on the generator but increases the risk of cavitation developing. Conversely, increasing the number of generator poles increases the cost of the generator.

Generator voltage will usually lie in the range 3—11 kV, at least for large machines. The final choice can be left relatively late in the decision process, but some early indication is needed so that approximate transformer and switch-gear sizes can be selected for inclusion in the barrage structure layout.

The number of turbines in each electrical group, and the number of turbines in each caisson (or other unit of construction) are linked to some extent. The minimum number is likely to be three turbines to a caisson, as discussed in Chapter 4, in which case the number of turbines in a group should logically be a multiple of 3. The same applies to other numbers. The number per group will be a function of available transformer capacities and the strength of the grid system, which will have to be able to accommodate the effects of a group shedding load. For the Severn barrage, early conclusions were that, based on 45 MW machines, there should be no more than ten turbines to a group, this being a round number and because the sudden loss of a total of 450 MW would be similar to the loss of one of the CEGB's large steam turbines, which are 500 MW and 660 MW. Clearly a small barrage with an installed capacity of 100 MW or less would not have a major effect on a national grid system but could affect the local distribution network.

The optimum generator capacity is not readily identified until much detailed work has been done. The economics of a barrage are not very sensitive to this factor, partly because the generators will represent around 10% of the cost of the barrage; so quite large changes in capacity will have little effect on total cost. What does seem clear is that the generators should be sized so that they restrict the output of the barrage for an hour or two during each spring tide, i.e. they should be somewhat smaller than necessary to generate all the power available on any tide. This is logical for two reasons. Firstly, the number of large spring tides each year is small, so generator and transmission capacity capable of using fully the maximum head available will be rarely used. Secondly, during these large tides, the water flow can be reduced when the generator capacity is reached, and the water 'saved' can be used later in the tidal cycle. A balance has to be struck, however, because water not used before the next flood tide starts to refill the basin is effectively wasted.

Having selected the diameter of the turbine runner early on, the selection of the number of turbines is one of the two fundamental decisions to be made. The other is the total area of sluice gates, discussed later. In essence, the task is to find the number such that the addition of another turbine increases the overall cost of the electricity produced by the barrage. This then identifies the optimum number in terms of the unit cost of electricity. Thereafter, turbines can be added until the cost of the electricity generated by the next turbine exceeds the cost of generation by alternative methods, this being the marginal cost approach. This assumes that there are no physical constraints imposed by the shape of the estuary at the proposed barrage site that limit the number of turbines.

These simple methods of identifying optima are complicated by a range of external factors, particularly the cost of borrowing money and the long term cost of fossil fuels, both of which will affect decisions regarding investment in other types of electricity generating plant, particularly thermal power stations. However, the various studies that have been done of tidal power schemes around the world have generally been coming to similar conclusions regarding optimum schemes, and the results can be used to speed up the study process.

Table 11.2, taken from Ref. 1987(13), summarises the main features of a number of tidal power schemes that have been studied around the world by different organisations, together with a number of sites in the UK for which preliminary studies (Refs. 1981(2), 1984(1) and 1986(1)) were carried out following the 1981 report of the Severn Barrage Committee. As can be seen, these cover a wide range of basin areas, tidal ranges, water depths and barrage lengths. Basin area and tidal range define the energy resource, while water depth, barrage length and, to some extent, tidal range define the volume and thus, approximately, the cost of the barrage. The cost data are at 1983 prices, this being the base date of Ref. 1984(1). To update costs to 1990, an approximate multiplier of 1.6 should be used.

In Fig. 11.1, a relationship is found between the total turbine runner area, represented by ND^2, and the energy resource (AR^2). Since the runner diameter (D) can be determined reasonably accurately from the minimum depth of water at the barrage site, this graph will give a good indication of the number of turbines required. Apart from two small schemes with small tidal ranges and which have been studied in only a preliminary fashion, the scheme that lies furthest from the line is the Rance barrage. It has been shown in Chapter 2 that more turbines are required to generate the same amount of electricity with two-way operation than one-way operation. Thus Rance, having more turbines, is consistent with this finding.

Fig. 11.2 relates the energy output per turbine to the mean tidal range. Very consistent results have been achieved around the world, the only exception being La Rance where the two-way turbines are equipped with relatively small generators and are less efficient than one-way machines, and so generate less electricity. The results for the two Severn barrage schemes are plotted twice, with and without the reductions in tidal range to seaward that have been predicted by large computer models of water movements. As may be expected, the post-barrage tide range gives the 'correct' energy output. Thus, having selected the size and number of turbines via Fig. 11.1, Fig. 11.2 allows the energy output of the barrage to be estimated with some confidence, bearing in mind that a barrage across a large estuary can be expected to reduce the natural tide range to seaward because it will be extracting a significant proportion of the energy of the tides.

Fig. 11.3 explores the relationship between the capacity of the generators and tidal range. The results are more varied than those on Fig. 11.2, partly because quite large changes in generator capacity have relatively little effect on energy output, as discussed earlier, and partly because the cost of the generators represents a relatively small proportion of total barrage costs. Again, the Rance generators are much smaller than would be needed if the barrage had been designed for one-way operation. The three barrages that have been studied in the Bay of Fundy have smaller generators than might be expected. This may be related to the very long transmission link that would be required to connect one of these barrages to suitable centres of demand. The cost of this link would be

Table 11.2 Summary of tidal schemes studied

	R Mean tidal range (m)	A Basin area (m²×10⁶)	L Barrage length (m)	H Max. water depth (m)	D Turbine dia. (m)	N Turbine no.	P Installed capacity (MW)	E Annual energy output (GWh)	C Capital cost (£M)	U Cost of energy (p/kWh)
1 Severn – Inner line	7·76 (7·0)	450	17 000	35	9·0	160	7200	12 900	6660	3·7
2 Severn – Outer line	7·2 (6·0)	1000	20 000	35	9·0	300	12 000	19 700	10 460	4·3
3 Morecambe Bay	6·3	350	16 600	30	9·0	80	3040	5400	3610	4·6
4 Solway Firth	5·64 (5·5)	860	30 000	28	9·0	180	5580	10 050	7480	4·9
5 Dee	5·95	90	9500	29	6·0	50	800	1250	1230	6·4
6 Humber	4·1	270	8300	29	9·0	60	1200	2010	2140	7·0
7 Wash	4·68 (4·45)	590	19 600	40	9·0	120	2760	4690	4860	7·2
8 Thames	4·2	190	9000	29	9·0	40	1120	1370	1740	8·3
9 Langstone	3·13	19	550	12	4·0	9	24	53	43	5·3
10 Padstowe	4·75	6	550	9	4·0	6	28	55	35	4·2
11 Hamford	3·0	11	3200	7	4·0	9	20	38	50	8·5
12 L. Etive	1·95	29	350	18	7·5	6	28	55	96	11·7
13 Cromarty	2·75	36	1350	27	7·5	6	47	100	176	11·8
14 Dovey	2·90	13	1300	11	4·0	9	20	45	50	7·2
15 L. Broom	3·15	7	500	27	7·5	3	29	42	90	13·9
16 Milford Haven	4·5	20	1150	27	7·5	6	96	180	270	10·0
17 Mersey	6·45	70	1750	25	7·6	27	620	1320	697	3·6
18 Fundy site A6	9·5	119	5410	42	7·5	53	1643	4530	2250	3·4
19 Fundy site A8	10·05	97	2560	39	7·5	37	1147	3183	1353	2·7
20 Fundy site B9	11·7	282	8000	42	7·5	106	4028	11 766	3875	2·2
21 Strangford Lough	3·1	144	1500	27	7·6	30	210	528	544	7·0
22 Garolim Bay (S. Korea)	4·8	100	1850	28	8·0	24	480	893	273	2·1
23 La Rance	8·0	22	750	23·5	5·35	24	240	544	—	3·4
24 Fundy site B9	11·7	282	8000	42	7·5	128	4864	14 004	4530	2·2

1. Costs at January 1983 price levels

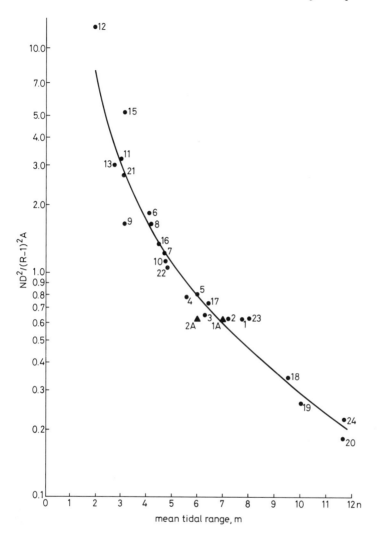

Fig. 11.1 Ratio of installed turbine capacity (ND^2) to energy resource $(R-1)^2A$ plotted against mean tidal range

significant and therefore the optimum generator size would be lower than normal.

The type of sluice gate is the next factor listed in Table 11.1 although this can be decided relatively early in the optimisation process. Because the gates are relatively expensive and can be expected to require regular maintenance, it will be logical to select a design of sluice which is as efficient as possible in terms of the flow it passes for a given head across it. The choices are discussed in Chapter 5. In essence, if there is enough water depth available, then the vertical

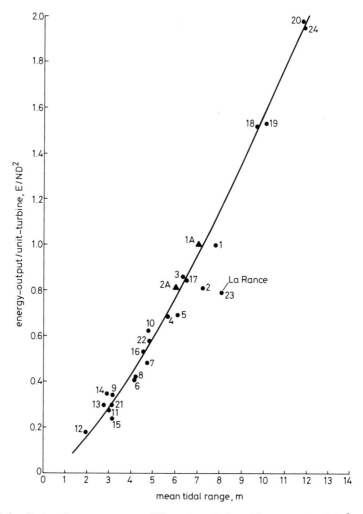

Fig. 11.2 Ratio of energy output (E) to installed turbine capacity (ND^2) plotted against mean tidal range

wheeled sluice gate is a good choice, located in a hydraulically efficient water passage. An alternative, especially for shallow water, is the radial gate.

The optimum total area of sluice gates is not simply related to the cost of energy produced by the barrage, although this is the main factor. A second factor is related to the important function played by the sluices during the construction of the barrage in providing a route for the tidal flows past the barrage site, thus reducing the differential heads and local velocities through the remaining gaps. This is discussed in Chapter 8. Although the overall costs of construction can be quantified and then compared with the number of sluices and the associated energy output, judgment has to be applied to the design as to

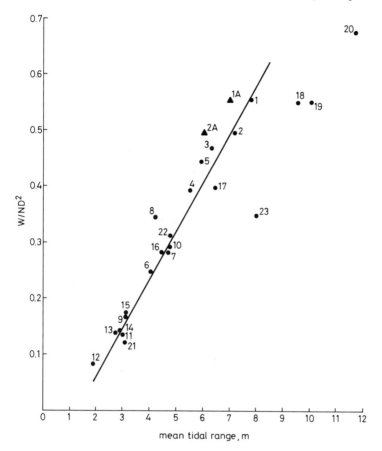

Fig. 11.3 Ratio of installed generator capacity (W) to installed turbine capacity (ND²) plotted against mean tidal range (R)

the maximum acceptable current velocities through gaps in the barrage, and especially to the length of time needed for critical current-dependent operations such as sinking caissons into position. If short 'current windows' can be tolerated, then there will be less need for additional sluice area.

The studies that have been carried out indicate that, for vertical gates in efficient passages, i.e. having a discharge coefficient C_d of about 1·8, the total area of the gates should be about double the gross area of the turbine runners, i.e. the area of the water passage at the runner, ignoring the space occupied by the runner and its hub. For other types of gate, the area should be increased in inverse proportion to their discharge coefficients.

The selection of the number and size(s) of ship locks is not a straightforward economic matter. If the estuary is used by a large number of ships, say 5000 or more movements (2500 ships entering and leaving) a year, then, although one lock may be enough to avoid significant delays, two locks are likely to be required in case one becomes unusable owing to an accident or routine maintenance. Clearly, at least one lock will have to be capable of handling the largest ship using the estuary. If further port developments are planned, these plans are likely to be based on still larger ships and this has to be taken into account.

If the barrage site is exposed to storms, substantial breakwaters will be needed, at least on the seaward side, to provide shelter for ships approaching the locks or waiting their turn to lock through. These will be expensive and their design will involve model tests and all the normal features of a major project.

Depending on the bathymetry of the seabed near the barrage site, the locks should be located where access will be available from the sea at all states of the tide, otherwise large ships will be delayed during their passages in each direction. If enough deep water is not available, then the alternative will be a dredged channel.

The locks will be required to be operational when the rest of the barrage begins to form a significant obstruction to tidal flows. This will be relatively early in the construction programme. Consequently the completion and commissioning of the ship locks will be a priority item and will adversely affect overall cash flows. Where a barrage site is located across a busy shipping route, the importance of ship locks in the economics of the tidal power scheme is not to be underestimated.

The 'working' parts of a barrage, namely the turbines in their power house, the sluices and the ship locks, would not normally be expected to occupy the full width of the estuary at the barrage site. Thus the remaining gaps have to be filled with some form of structure which will both provide an access route and close the gaps to tidal flow. This will be either more caissons or embankments. The caissons could be relatively simple concrete or steel boxes, floated into position and then filled with sand or other ballast to make them stable against the differences in water level resulting from the barrage working. These are referred to as plain caissons.

Plain caissons require two conditions to be practical: enough water depth to enable them to be floated into position, and adequate foundations. This simple statement covers some complex questions. To start with, plain caissons would, logically, not be placed until after most of the rest of the barrage components, particularly the ship locks and turbine caissons, had been installed. As discussed in Chapter 8, this means that the time during which slack water is available for positioning each caisson and ballasting it down is short, and so the risk of problems and delays is increased.

Because the sides of estuaries usually contain the bulk of the sediments present, the availability of foundations of adequate strength to support plain caissons must be checked before any decision to use them is taken. Following from this, the likelihood of wholesale scouring of the sediments as current strengths increase through the remaining gaps also has to be considered.

Embankments can and have been designed to be built on poor foundations, and do not require a minimum water depth to be placed. In addition, the final stages of 'closure' are similar to the final stages of diversion of a large river to enable a dam to be built, with which much experience has been gained around the world. Thus embankments are the logical choice for the shallowest parts of the estuary and also form a 'fall-back' solution for other parts of the site. Cost estimates for plain caissons and embankment prepared for the Severn barrage show plain caissons to be cheaper for water depths greater than about 15 m. A secondary issue is wave reflection; the vertical faces of plain caissons will reflect waves efficiently, whereas the sloping faces of embankments will cause waves to break and their energy to be dissipated. Thus the areas near embankments will be quieter in terms of the wave climate and more comfortable for sailors. The importance of this will vary from site to site.

The results of Figs. 11.1–11.3 are combined in Fig. 11.4, which shows the relationship between a cost function, comprising essentially the volume of the barrage divided by the energy potential of the site, and the estimated unit cost of energy, at 1983 prices, calculated by dividing the discounted total capital and operating cost of the barrage (at 5% real rate of interest) by the total discounted energy output over the lifetime of the barrage. A reasonable relationship has been found, bearing in mind the fact that the cost function relies on the volume of the barrage and does not take into account the installed capacity.

The penultimate item in Table 11.1 is the cost of alternative forms of electricity generation. This affects the value of the electricity from tidal power, which is discussed later, but also affects the process of identifying the optimum configuration of the barrage. In essence, having identified the numbers of turbines and sluices, and generator capacity, which result in the lowest unit cost of electricity from a barrage, the barrage's position in relation to other forms of generating electricity can be checked.

If the barrage's cost of electricity is less than that of the next alternative, for example electricity from coal, then additional sluices and turbines can be added to the barrage until the cost of the additional electricity produced reaches that of the alternative, noting that the addition of working parts will reduce the cost of the non-working parts, namely the embankments or plain caissons. Although this approach is valid, in practice the marginal cost of electricity from additional turbines tends to be relatively high because the width of the deep water channel is usually restricted. The marginal cost of the energy from

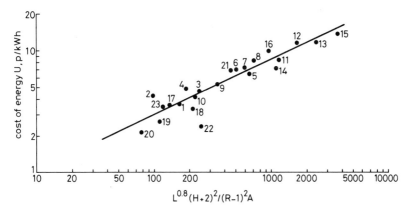

Fig. 11.4 Cost of energy (U) plotted against energy cost function (C)

additional sluices (which allow more water into the basin) rises less steeply because sluices can be readily designed to suit whatever water depth is available.

Another factor which may be important in affecting the marginal cost of electricity is the transmission link between the barrage and the grid. For a given transmission voltage, there will be a limit to the power that can be transmitted along a single circuit. Thus the addition of another circuit, due to the addition of more turbines, will cause a steep change in cost.

The last item in Table 11.1, interest rates, is perhaps the most important in determining the cost of electricity generated by a tidal power barrage. This can be demonstrated quite simply. Firstly, no fuel costs are involved, and fuel costs are a major factor in the cost of electricity from thermal power stations. Secondly, the costs of operating and maintaining a tidal barrage, assuming it has been designed and built properly and with a view to achieving a long working life, have been estimated to represent less than 1% of the capital cost per year. Thus the interest payable on the capital cost of the barrage, which is unlikely to be less than 5% per year in real terms, makes up at least 80% of the cost of the electricity generated until the capital debt has been reduced significantly.

As well as affecting the financing of the capital debt, interest rates affect the optimisation process in several ways. Most obviously, the time taken for construction should decrease as the interest rate rises, in order to reduce the total amount of interest charged on borrowings before the barrage starts to generate electricity and pay for itself. Reducing the time for construction will require additional resources in men and plant, and may involve greater risks because activities have to take place concurrently instead of sequentially. This will be particularly important for the programme for the 'non-working' parts of the barrage, the embankments and plain caissons, which should not be built until the expensive working parts have been safely installed.

The fundamental affect of interest rates on the cost of electricity from a tidal power barrage is illustrated by Fig. 11.5, taken from Ref. 1981(1). Thermal power stations are much less sensitive to interest rates because interest payments on capital costs are a smaller proportion of total costs, fuel costs dominating.

11.3 Value of electricity from tidal power

The value of electricity can be quite different from its cost, and consideration of this introduces a range of aspects not connected directly with either engineering or the interest rate payable on capital.

The value of electricity to a person will depend very much on personal circumstances. In the western countries, most people have come to depend on electricity being constantly available at a reasonable price for much of their existence: cooking, lighting, some part of their heating and/or air conditioning, washing, entertainment, communications and so forth. A large increase in the cost of electricity would be needed to make significant changes to the lives of many of these people, because they could not easily manage without electricity and they see it as having great value. On the other hand, there are people in the

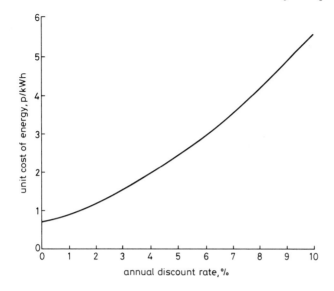

Fig. 11.5 Variation of unit cost of energy with discount rate

remoter parts of western countries who, in their normal lives, manage very well without electricity, except perhaps for batteries for radios or torches. For these, and for most of the populations of the world, electricity is not an essential ingredient of life and therefore has relatively little value.

If electricity has little value, there will be little chance of an estuary or inlet with a good shape and large tidal range being suitable for development. As an example, there are several large inlets on the north-west coast of Australia which could be highly suitable for tidal power development (Chapter 12). However, in the absence of any sizeable local demand, there is no foreseeable likelihood of their being developed.

Assuming that a potential tidal power site is located in a developed country where electricity is central to the standard of living, the question is whether the site should be developed or whether the money would be better invested in another type of power station, e.g. coal-fired, nuclear, hydro, wind turbines or whatever. A secondary question is who should develop the site: a public utility or a private organisation?

The answer to the first question depends very much on whether the other options are available in the country concerned. In general, if a country has large reserves of coal, then these have formed an important source of fuel for thermal power stations. The UK is a prime example, generating around 80% of its electricity from coal. If there are sites suitable for large scale hydro-electric power, then these are a prime source of electricity. Norway, Switzerland, Canada and Brazil are examples of countries which generate much of their electricity from water power.

Without reserves of coal, gas or sites for hydro-electric power, the choice turns to nuclear power stations. France generates about 70% of its electricity from nuclear power, and much of the rest from low-head hydro stations on its main rivers, particularly the Rhone. After the Second World War, uncertainty about the future sources of electricity led to the French government deciding to build the Rance tidal power scheme as a pilot for a much larger scheme to seaward. The latter, often referred to as the Chaussey scheme because it could be built out to the Iles de Chaussey, has been studied intermittently over the years, with emphasis on generating electricity as continuously as possible by using two basins (Ref. 1987(6)). This suggests that, had nuclear power not been available, France may have developed tidal power on a massive scale and, by doing so, saved imports of large quantities of coal or oil.

Turning to the UK, it is instructive to recall how the Severn barrage has been viewed over the years. In 1927, the government made three decisions concerning the electricity supply system. It decided that the standard frequency would be 50 Hz and that a national grid should be built. It also decided to build the Severn barrage, but, because the best site had not been identified, appointed a committee under Lord Brabazon to study the problem. The committee, the Severn Barrage Committee, reported in 1933 (Ref. 1933(1)), concluding that the barrage should be located at the English Stones and have a total capacity of 800 MW. Because the national grid was not yet developed, the committee also decided that the intermittent output of the barrage would have to be smoothed by an associated pumped storage scheme. As a result of this increase in cost, and the losses in the pumped storage cycle, the overall cost of electricity would not have been competitive with that from coal stations. Fears for employment in the south Wales coal mines led to the scheme being deferred.

The Severn barrage was reassessed in 1945 (Refs. 1945(1) and 1949(1)), by which time the national grid had developed enough to enable the power to be absorbed without the need for associated pumping storage. The cost of electricity was calculated at 0.115p/kWh at 3% real rate of interest, lower than that from existing coal stations but more than that from planned new stations. Thus, again, the scheme did not proceed. However, meanwhile, in Scotland development of hydro-electric projects started, primarily not because they would generate electricity more cheaply than coal stations, but because they would develop a resource of renewable energy and create employment. Now, with their cost repaid they are generating the cheapest electricity in the UK and, given adequate maintenance, will continue to do so for the foreseeable future.

The second Severn Barrage Committee was appointed in 1978 and reported in 1981 (Ref. 1981(1)). The Committee concluded that the Severn barrage, now located near the Holm islands and with an installed capacity of 7200 MW instead of 800 MW, would produce electricity at lower cost than new coal stations but higher than nuclear stations. The amount of coal saved would be about 6 million tonnes a year. However, much work remained to be done on engineering and environmental aspects before a fully informed decision could be taken to build the barrage, which the Committee recommended should proceed at a cost of about £20 million.

At this point, it can be seen that the decision whether or not to build the Severn barrage has been linked with the view taken at the time as to the future of coal-fired stations and, latterly, nuclear stations (e.g. Ref. 1983(4)). If the

UK did not have large reserves of coal, then it would seem that the Severn barrage would have been built and, with zero fuel costs like hydro schemes, would eventually be generating the lowest cost electricity (as long as its basin did not silt up significantly).

All the evaluations carried out up to the report of the second Severn Barrage Committee were based on a discount rate, effectively the real interest rate above inflation, of 5%. This is slightly higher than the historic rate of return from low-risk investments, which is about 3%, and was the rate recommended at the time by the Treasury for public sector projects. It is worth noting in passing that some of the large public utilities do not return even 5% on their invested capital. Until 1987, CEGB was achieving less than 3%, as were the Water Authorities. As a result, consumers could be said to be paying much less than they should be, and this is one reason why electricity charges can be expected to rise following the privatisation of the CEGB.

The process of discounting is used to enable alternative investments to be compared on a similar financial basis. In layman's terms, any future item of expenditure or income can be given a present value, i.e. equated to a sum of money which, placed in a bank paying interest at the rate selected for the discounting process, will increase over the years to equal the future item of expenditure or income. If the discount rate is 5%, then £100 placed in a bank now will pay for something costing £105 in a year's time, or something costing $£100 \times (1 \cdot 05)^{100} = £13\ 150$ in 100 years' time. If the discount rate is 10%, then the first figure increases to £110 but the second figure increases to £1 378 061. In the same way, present values can be attributed to future income. Thus an income of £10 next year will have a present value of £9·52 at 5% discount rate, while £10 in a hundred years' time will have a present value of £0·076. With a discount rate of 10%, the latter figure drops to £0·00073.

Once the capital cost and construction period of a tidal power scheme have been established it is a straightforward matter, having selected a likely completion date, to calculate the discounted cost to any selected base date. In the same way, the energy generated during the expected life of the barrage, together with annual maintenance costs and foreseeable major items of expenditure to replace plant, can be discounted back to the same base date to obtain total discounted costs and total discounted energy output. Dividing total costs by total energy output will give the discounted cost of energy over the life of the barrage. This can then be compared with the cost from other types of electricity generation.

The examples of present values in the previous paragraphs have been selected to demonstrate the sensitivity of the economics of long-term investments to the chosen discount rate. Indeed, this effectively dominates the decision process concerning power stations because a high discount rate will favour plant with a low initial cost, high running costs and short life. For example, the present value of replacing a power station in 25 years time at 5% or 10% discount rates will be 29·5% or 9·2%, respectively, of their capital cost. In the same way, a unit of electricity worth say 4p today is given a present value of either 1·18p (5%) or 0·37p (10%) in 25 years time.

The advantages of tidal power include the fact that the tides are a truly renewable and predictable source of energy and that, given careful design and construction, a tidal power scheme should have a long working life—La Rance barrage is in excellent condition after its first 22 years of operation. Unfortunately, the discounting process means that no sensible value can be put on these advantages, because a benefit (electricity) produced in 25, 50 or 100 years time has little present value, whereas the real value of electricity, in a future where oil is in short supply and concern grows about the 'greenhouse' effects of large discharges of CO_2 to the atmosphere, could be large.

The UK Department of Energy carried out a review of the prospects of the various renewable energy sources in 1987 (Ref. 1987(7)). This showed that wind power was the most promising source which could be developed on a large scale (hydro in the UK is not a large resource), with tidal power in the second rank. A wide-ranging debate on the subject was held in the House of Lords (Ref. 1987(1)).

Chapter 12

Tidal power resources around the world

12.1 Introduction

This chapter summarises the locations around the world where the combination of large tidal range and suitably indented coastline are of interest for tidal power development. For this purpose, based on the results discussed in the previous chapter, it has been assumed that the mean tidal range $(2 \times M2)$ should be about 5 m or more, this giving a reasonable chance of a unit cost of electricity of 6p/kWh or less (assuming 5% discount rate and UK construction costs). For any site with adequate tidal range and suitable bathymetry, the likely size of the resource and the main components of a barrage to suit the site can be estimated using the graphs presented in the previous chapter. The tidal range and the area enclosed define the energy available, while the length of the barrage and the maximum water depth indicate the volume of the barrage and thus its cost. Although tidal range is the most important criterion when assessing a site, the area of the enclosed basin should be as large as possible in proportion to the length and depth of the barrage, in order to maximise the ratio of the energy generated to the cost of the barrage.

The lengths of coastline around the world where the mean tidal range exceeds 5 m are shown in Fig. 12.1. Also shown are some locations where the mean tidal range is at least 4 m and the coastline could be favourable for tidal power development. Each area is discussed separately in the following Sections.

12.2 Alaska

On the south coast of Alaska lies Cook inlet and the town of Anchorage (Fig. 12.2). Here, at Kinick Arm, the mean tidal range reaches 7·14 m. The tides have a moderate semi-diurnal inequality, with successive low tides varying more than high tides (Fig. 12.3). This inequality would have little effect on the operation of a barrage because the new low water level behind the barrage would be at about mean sea level.

Preliminary studies have been carried out of this site (Ref. 1981(19)). Unfortunately, the inlet is shallow at low tide, making it difficult for turbines to be located deep enough for satisfactory operation. Another problem is that the climate is severe, with ice presenting difficulties in winter and in the spring thaw.

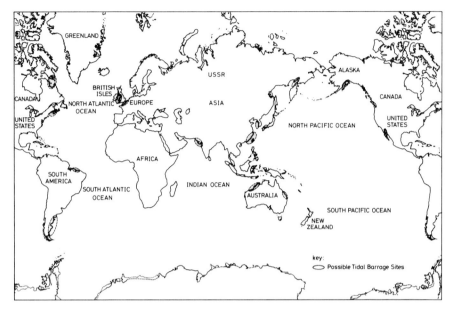

Fig. 12.1   Locations of possible sites for tidal barrages

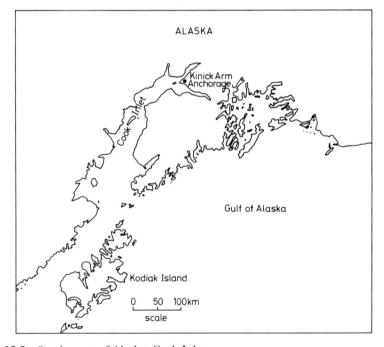

Fig. 12.2   South coast of Alaska: Cook Inlet

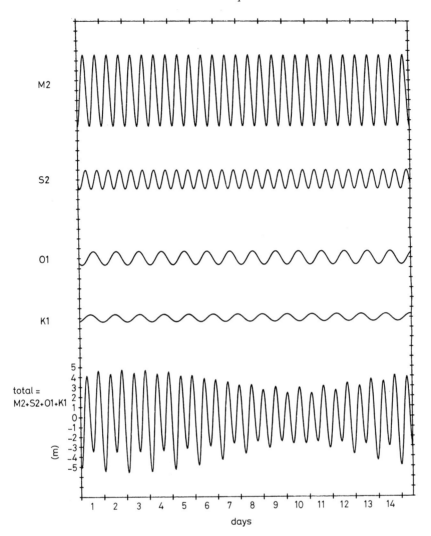

Fig. 12.3 Tides at Kinick Arm, Anchorage

12.3 Argentina and Chile

About 1000 km south of Buenos Aires there are three large bays, each of which represents a very large tidal energy resource (Fig. 12.4). Their relevant features are summarised in Table 12.1.

For the entrance to the Golfo Nuevo, Admiralty Tide Tables include data for Punta Ninfas, which is on the south side. As can be seen from the predicted unit cost of energy, this site suffers from a mean tidal range below the target figure of

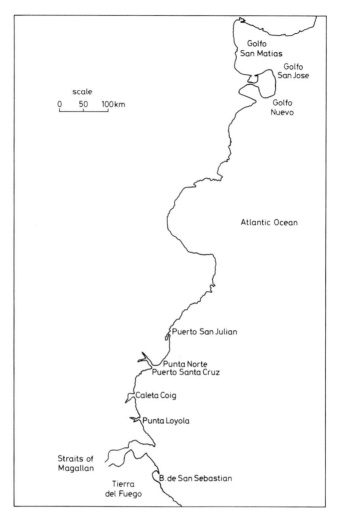

Fig. 12.4 South Argentina: possible barrage locations

5 m, and from a large depth of water which would make construction difficult and expensive.

Admiralty tide data for Golfo San Matias show the mean tidal range increasing from 4·22 m on the south side of the entrance to 5·9 m on the north shore and then 6·28 m in the north-west corner. It is not obvious what the mean tidal range would be at a barrage running due north across the mouth, partly because such a large scheme could be expected to reduce the natural range somewhat. The range given in Table 12.1 is therefore an estimate only. On this basis, the scheme would generate electricity equivalent to the burning of about 70×10^6 t of coal a year, but at a unit cost which is rather high. In addition, the absorption of this amount of variable power by the rest of the power supply

Table 12.1 Summary of tidal resources south of Buenos Aires

Site	Area (km^2)	Max. depth (m)	Length (m)	Mean range (m)	Turbines (no. × dia.)	Annual energy (TWh)	Unit cost (p/kWh)
Golfo Nuevo	2260	90	16500	3·66	415 × 9	12·8	9
Golfo San Matias	15050	90	96000	5	327 × 9	160	5·1
Golfo San Jose	788	25	7000	5·78	270 × 7·5	10·9	2·1

system would present formidable problems, even allowing for the fact that Argentina has large hydro power stations which could be cycled more easily than thermal plant.

The third scheme listed above, located at the mouth of the Golfo San Jose, is the most interesting of the three. At 6000 MW, its size is more practical and the predicted unit cost of energy is one of the lowest of all large sites around the world. The tidal range given is that for San Roman, just inside the east side of the entrance. The similar tidal range in the main bay outside suggests that a barrage here would not reduce the tidal range at the site significantly. The bay appears to have been formed by the erosion of a gap through the coast followed by land occupying what is now the bay; certainly, Admiralty Chart 3067 shows the entrance to be shallow compared with the approaches from seaward and landward, suggesting that there is a bar of hard ground across the entrance. The maximum depth is given as 7 fathoms (12·8 m)) below chart datum, i.e. about 21 m at high water springs. For the estimate of costs, this has been increased to 24 m to allow for inaccuracies in the small scale chart.

For all these three sites, there is no evidence on the charts to suggest that sedimentation of the basins would be a problem.

The southern end of the Argentine coast and the east coast of Tierra del Fuego, which belongs to Chile, have large tidal ranges, peaking at a mean range of 7.86 m in the estuary in the Rio Coig near Rio Gallegos (Fig. 12.5). This is comparable with the range in the Severn estuary. As well as this estuary, there are three other estuaries of interest: San Julian, Rio Santa Cruz and Rio Gallegos. Their main features are summarised in Table 12.2. Because all have narrow entrances and large tidal ranges, they appear to be exceptionally attractive sites with cost functions in the range 22–60 (see previous chapter). Thus their predicted unit costs of energy are low. Three of the sites are shallow, with the chart indicating extensive intertidal flats in the estuaries and sandbanks outside. This suggests that these estuaries would suffer from siltation if developed. The fourth estuary, that of the Rio Santa Cruz, is deeper, 25 m being indicated at the barrage site on the chart. This would be more practical, enough for the largest turbines available. Sandbanks offshore may present a threat of sedimentation in the basin.

The Straits of Magellan (Entrecho de Magellan) and the complex system of waterways through to the west coast of Chile are fascinating to study, with a wonderful mixture of Spanish and English names to stir the imagination: Seno Christmas, Punta Anxious, Punta Hope. The entrance to the Straits has a mean tidal range of 6·6 m. About 70 km inland (Fig. 12.6) are narrows called Prima Angustura (First Narrows). Here the mean tidal range is 5·6 m. About 50 km further are second narrows, Segunda Angustura, where the mean range is not

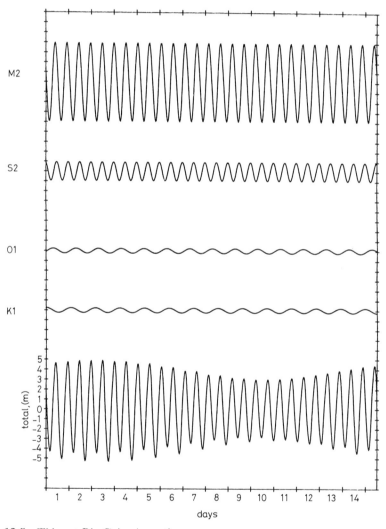

Fig. 12.5   Tides at Rio Coig, Argentina

Table 12.2 Estuaries in the south Argentine coast and the east coast of Tierra del Fuego

Site	Area (km²)	Max. depth (m)	Length (m)	Mean range (m)	Turbines (no. × dia.)	Annual energy (TWh)	Unit cost (p/kWh)
Rivers							
San Julian	77	13	810	5·66	40 × 6	1·04	1·8
Rio Santa Cruz	215	32	2070	7·48	60 × 9	5·05	2·3
Rio Coig	46	12	1800	7·86	30 × 6	0·61	1·9
Rio Gallegos	140	12	3400	7·46	85 × 6	3·27	1·6
Straits of Magellan							
Entrance + Prima Ang.	2230	75	28 600 + 3900	6·6	560 × 9	40	5·2
Entrance + Segunda Ang.	3340	75	28 600 + 6900	6·6	840 × 9	59	4·5
Prima Ang. + Segunda Ang	1110	52	3900 + 6900	5·6	260 × 9	14	4·0
Bahia San Sebastian	580	30	19 300	6·5	145 × 9	10	3·8

given in the tide tables but may be about 3 m. Beyond these, the mean range drops to about 1 m and remains at this or less through the system of channels which connects through to the west coast and the Pacific Ocean. This pattern of tides decreasing from the mouth of the straits inland is unusual; often in long estuaries or inlets which have large tidal ranges, the range increases inland as a result of resonance.

In theory, it should be possible to build a tidal power barrage across the mouth of the straits and an impermeable barrage at Segunda Angustura, the latter to stop tides 'escaping' at the west end. Alternatively, a smaller impermeable barrage could be built at Primera Angustura. Yet again, a power generating barrage could be built at Primera Angustura in combination with the impermeable barrier at Segunda Angustura. A computer model of the tides would be needed before the performance of these combinations could be forecast with confidence. Meanwhile, assuming that the tidal range to seaward of either of the postulated power barrage sites would be largely unchanged, because the driving tide is that at the entrance to the straits, the principal features of the barrages are listed in Table 12.2. It is noteworthy that the Straits are the site of much offshore oil production, so interference with this would have to be avoided.

The last very large tidal power resource is Bahia San Sebastion, on the east coast of Tierra del Fuego. Here the mean tide range is still 6.6 m. The logical site for a barrage would be straight across the entrance, but the Admiralty chart shows a local deepening from a typical depth of around 22 m to double this at the northern end, Punta de Arenas. By kinking the barrage line, this deep spot could, apparently, be avoided. The resulting features of a barrage are included in Table 12.2.

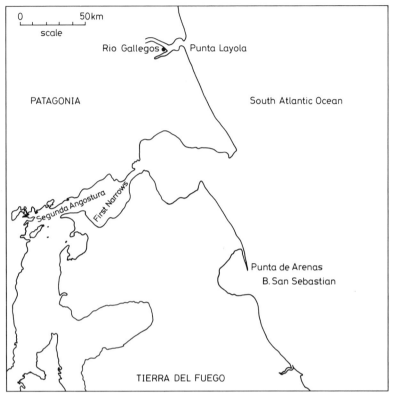

Fig. 12.6   Straits of Magellan

As can be seen, the river estuaries, with their narrow and shallow entrances and moderate size, appear worthy of more detailed evaluation to check their economic and environmental feasibility. Of the large sites, that with the lowest unit cost of electricity would be Bahia San Sebastion. However, like all the sites at the southern end of the continent, it is very remote from centres of demand from electricity and from suitable construction resources. It would also interfere with oil-related facilities on the south shore, and there may be problems associated with mobile sediments—the chart suggests extensive sandbanks either side of the seaward side of the entrance. Thus these large sites are of theoretical interest only for the foreseeable future.

12.4 Australia

Around most of the coast of Australia the tidal range is generally too small to be of interest. On the east coast of Queensland, between Mackay and Rockhampton, the mean tide range exceeds 3 m and peaks at 4·84 m at McEwin Islet in Broad Sound (Fig. 12.7). However, the width of the entrance of

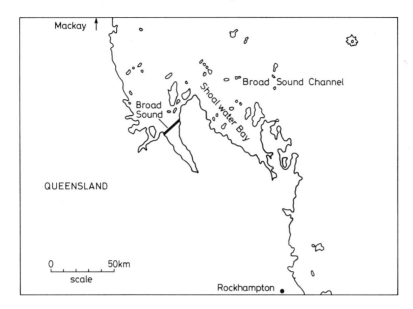

Fig. 12.7 East coast Australia (Queensland)

Broad Sound, as may be expected, is large in proportion to the enclosed area, so this site is of little interest.

Much more interesting possibilities exist on the north coast of Western Australia, south-west of Darwin, where there are several large bays or inlets and the mean tidal range reaches 6·5 m. Fig. 12.8 shows the main sites.

The largest is King Sound, where the mean range increases from 4·34 m at Sunday Island at its mouth to 6·14 m at Derby, at its head. The parametric method has been applied to two lines and the results are shown in Table 12.3. These show that King Sound would be an enormous resource which could be developed at a unit cost of about 5p/kWh. However, the site's remoteness from any centres of demand makes its development impractical.

The next two sites are Secure Bay and Walcott Inlet, and these have already been the subject of two preliminary studies (Refs. 1963(1) and 1976(1)). The mean tidal range at Shale Island, off the mouth of these inlets, is 6·52 m, which is large enough to be attractive. These two sites also have entrances which are exceptionally narrow (but deep) in proportion to the enclosed areas. Although beneficial in reducing the length of the barrage, the tidal currents are very fast and would present severe problems during construction (Fig. 12.9). In Ref. 1976(1) an outline design was developed which was based on caissons which had a large proportion of their area open to tidal flows into which flap gates could be installed, when the barrage was ready to be commissioned, to form self-operating sluice gates. The sluice design is discussed in Chapter 5.

Either, or both, of these inlets could be developed as ebb generation schemes. Using the parametric method, the results in Table 12.3 are obtained.

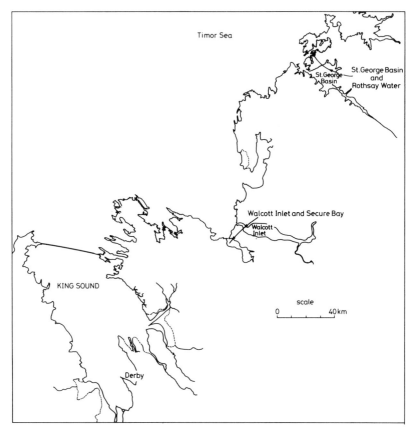

Fig. 12.8   Western Australia: possible barrage locations

As can be seen, Secure Bay appears to be the more economical, largely because it has a more manageable depth.

Because these sites are remote, their varying output could not be absorbed readily. Thus other options have been considered which would result in a smoother output. Because of the proximity of the two inlets, one possibility

Table 12.3   Sites on north coast of W. Australia

Site	Area (km²)	Max. depth (m)	Length (m)	Mean range (m)	Turbines (no. × dia.)	Annual energy (TWh)	Unit cost (p/kWh)
Secure Bay	140	50	1300	7	37 × 9	2·9	3·6
Walcott Inlet	260	75	2500	7	70 × 9	5·4	5·1

Fig. 12.9 Walcott Inlet, Western Australia
(Photo: P. Ackers)

would be to connect them together by deepening a connecting channel and then locating the turbines in the channel. By using Secure Bay as the high level basin, filled by inward-pointing sluices at high tide, and Walcott Inlet as the low basin, emptied by sluices at low water, the turbines could generate a modest amount of electricity continuously from the difference in level between the two basins. Unfortunately, the $S2$ tidal constituent, which governs the spring–neap cycle, is a larger proportion of the mean tidal constituent, $M2$, than normal. At Shale Island $M2$ is 3·26 m and $S2$ is 1·98 m. Thus the spring tide range is 10·48 m but the neap range is only 2·58 m. Fig. 12.10 shows the four main constituents and the resulting spring–neap cycle. The neap range is so small that it proved impossible to generate any useful power through neaps without an associated inland pumped storage scheme. The combination would therefore not be economic.

St. George's Basin and Rothsay Water, about 100 km to the north, also offer a wide range of possibilities for tidal power. These sites have been the subject of preliminary studies in connection with possible mining of bauxite at Mitchell

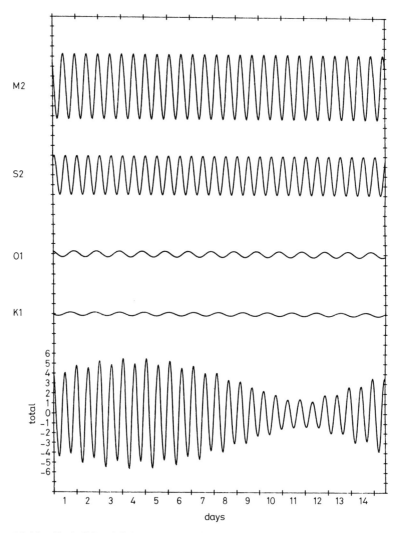

Fig. 12.10   Shale Island, Western Australia

Plateau to the north. To extract aluminium from bauxite, two stages are necessary. First the bauxite is converted to alumina by adding caustic soda and heat. Then the alumina is converted to aluminium by an electrolitic process. The usual method, the Hall–Heroult process, requires 13 000—20 000 kWh of electricity to produce one tonne of aluminium from about 1·9 t of alumina. The process is continuous, with severe problems being met with the smelters if there is a power failure of more than about an hour. Since the indigenous hydro-electric potential is limited to the Mitchell River, estimated at about 6 MW, the possibility of producing electricity from a two-basin scheme was investigated. Such a scheme could comprise one of the following alternatives:

- a normal ebb generation barrage in one inlet, and a flood generation barrage in another
- one set of turbines located between the two basins, with one basin kept at high level by inward facing sluices and the other kept low by outward facing sluices
- a normal ebb generation barrage in one basin, with the other used as a pumped storage scheme with reversible turbines
- a normal ebb generation barrage in one basin, with an inland pumped storage scheme.

These options have all been considered, with the aim of producing a continuous output of about 200 MW, enough to supply a smelter producing about 100 000 t of aluminium a year.

Fig. 12.11 shows the pattern of tides and Fig. 12.12 is a map showing alternative locations for barrages near the entrance to St. George Basin and Rothsay Water. As a simple ebb generation scheme each site would have the features detailed in Table 12.4.

In all cases, the large depth of water and modest tidal range offset the attractiveness of the short barrage lengths.

Fig. 12.13 shows the performance of St. George Basin as an ebb generation scheme with the number of turbines reduced in order to reduce the cost of the scheme and concentrate on generating more continuous power than normal, based on the main constituents M2 and S2 only. Thus installed capacity has been reduced from 50 turbines to 10, and the total capacity reduced to 400 MW instead of 1000 MW. As can be seen, the drop in water level each tide is much less than normal, but the small neap tides result in no power being generated.

Also shown in Fig. 12.13 is the effect that a separate pumped storage scheme would have in tidying the output. A storage scheme with an installed capacity of 330 MW pumping and 70 MW generating would result in continuous output of 70 MW. The total installed capacity of 730 MW for a continuous output of 70 MW does not appear attractive.

Fig. 12.14 shows the performance of the two basins when the smaller, Rothsay Water, is used for ebb generation, and St. George Basin for flood generation. In this case, the smaller basin should be used for ebb generation because then its area will be a maximum, while the flood generation's basin effective area will be reduced.

In this case, the opportunity has been taken to site a single set of turbines in a channel between the two basins (Fig. 12.15). Thus the installed capacity is reduced to only five turbines, 83 MW total, with the works at the entrances to each bay being only sluices and embankments.

As can be seen from Fig. 12.14, the scheme works well through spring tides and down to mid-range tides. The small neap range proves insurmountable for continuous generation. Additional turbine capacity and associated pumped storage capacity could produce continuous power with a much smaller total capacity than for the single basin scheme discussed above, but costs would still rise as a result.

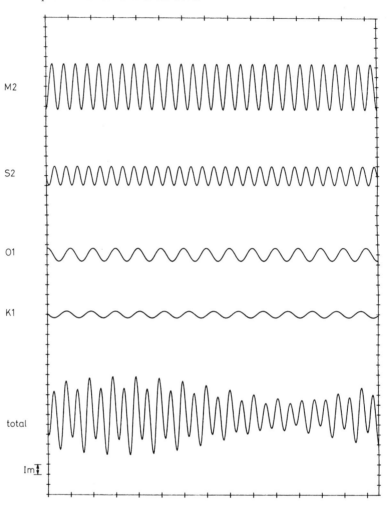

Fig. 12.11   Tides near Rothsay Water

Fig. 12.16 shows the same scheme but with the installed capacity reduced to 33 MW in an attempt to bridge the gap at neaps. This time, the large basin storage plays a key role through neaps, staying below the level of low tide at neaps. For nine days out of the 14.5 day cycle, power is maintained at 33 MW. During neaps, power is still continuous but slips to a minimum of about 10 MW. Needless to say, the engineering works at the mouth of each bay would still be extensive and could not justify a scheme with an output of only 33 MW.

Fig. 12.17 shows the performance of the two basin scheme acting as its own pumped storage. With this arrangement, Rothsay Water would have reversible pump turbines, which would be used to absorb excess power from the St. George Basin turbines during spring tides by raising the water level in Rothsay

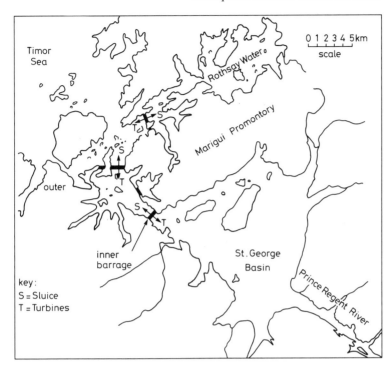

Fig. 12.12 Western Australia, St George Basin and Rothsay Water: single basin schemes

Water much higher than normal. During neap tides, the Rothsay Water turbines would be operated nearly continuously, reducing output only for the short periods when St. George Basin's turbines were operating.

With this combination, continuous power of 225 MW would be generated, thus meeting the target. However, the total installed capacity would be about 2000 MW, which would make the scheme expensive. The environmental effects

Table 12.4 Sites for barrages at St. George Basin and Rothsay Water

Site	Area (km^2)	Max. depth (m)	Length (m)	Mean range (m)	Turbines (no. × dia.)	Annual energy (TWh)	Unit cost (p/kWh)
St. George inner	227	55	1000	4.25	43 × 9	1·60	5·2
St. George outer	257	65	1500 + 350	4.25	50 × 9	1·87	7·5
Rothsay Water	75	50	800	4.25	15 × 9	0·55	7·5

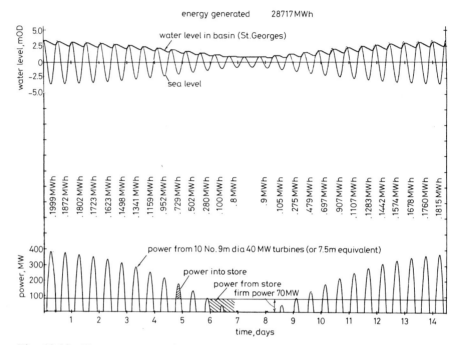

Fig. 12.13   Energy output and water levels, single basin and inland PS

of the water level in Rothsay Water varying between about + 10 m and + 4 m over a 14 day cycle have not been considered. The hinterland is an Aboriginal Reserve.

Fig. 12.18 shows how the variable output of a 250 MW interbasin scheme could be smoothed by an inland pumped storage scheme of about 150 MW capacity, to give a firm power of 150 MW with a total installed capacity of only about 400 MW, much less than with the pumped storage element built in to a tidal scheme.

To conclude, the north-west coast of Australia has some outstanding tidal power sites, with large basins, large spring tides and narrow entrances. To offset these advantages, the neap tides are relatively small, the entrances are deep and there is no practical method of absorbing the intermittent output from a tidal scheme without very long and costly transmission links. Methods of smoothing the output of a tidal scheme have been identified, but these would add significantly to the cost of what would already be relatively expensive electricity.

12.5 Brazil

The coastline of the territories of Amapa, Para and Maranhao, on the north coast of Brazil, has numerous bays and inlets (Fig. 12.19). The mean tidal range along this coast is about 4 m, reaching 4.37 m at Sao Luis and 4.44 m at Itaqui,

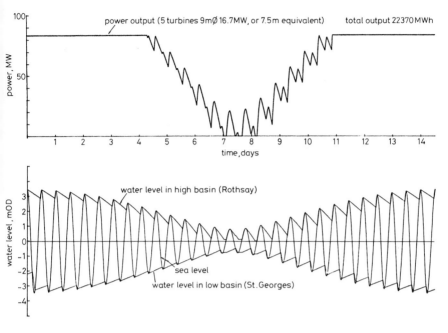

Fig. 12.14   Energy output and water levels: interbasin scheme

according to Admiralty Tide Tables. Sondotecnica, on behalf of Electrobras, the state electricity generating organisation, has carried out a preliminary survey of the tidal resource along this coast (Ref. 1980(13)) and lists the mean tide range near Amapa, to the west of the Amazon delta, as 8 m. This is exceptionally large and is not confirmed in Admiralty tide tables. Along the rest of the coast, the mean range of about 4 m is less than the target figure for reasonable prospects of 5 m. Sondotecnica's report lists 42 possible sites, ranging in area from 616 to 15 km^2, and in installed capacity from 4912 to 60 MW. The total resource is estimated at about 27000 MW, producing about 72 TWh/year. Apparently, twelve sites have been short-listed, with capital costs at August 1980 price levels in the range \$1000–1500/kW. Unit costs of electricity were estimated at \$0.0405–0.053/kWh, or around 2–3p/kWh. Using the parametric method of assessment for the largest site listed, the Bay of Turiaco, gives the following:

Depth of water at MHWS	10 m approx
Turbine diameter	5 m say
Number of turbines	440
Installed capacity	4100 MW
Annual energy output	8200 GWh
Unit cost of energy	2·3 p/kWh (3·5p/kWh from minimum capital cost of £1000/kW installed)

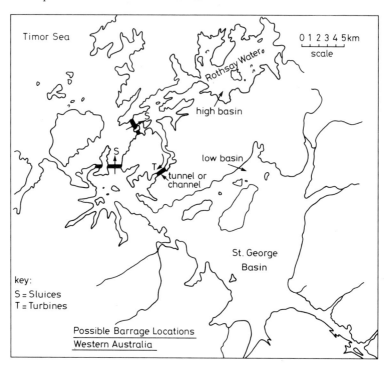

Fig. 12.15   St. George Basin and Rothsay Water: interbasin scheme

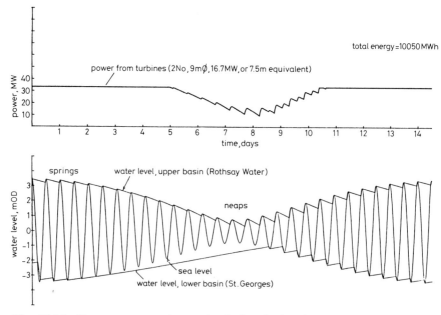

Fig. 12.16   Energy output and water levels: interbasin scheme

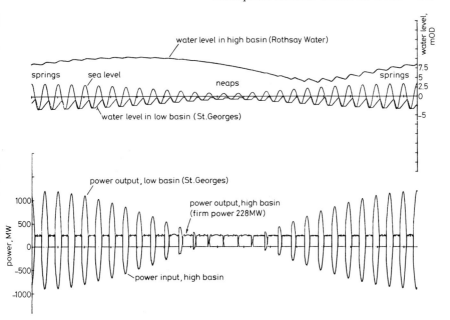

Fig. 12.17 Energy output and water levels: two-basin pumped storage scheme

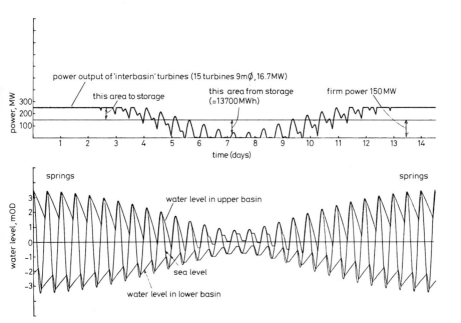

Fig. 12.18 Interbasin scheme with inland high-head pumped storage: King George's Basin and Rothsay Water

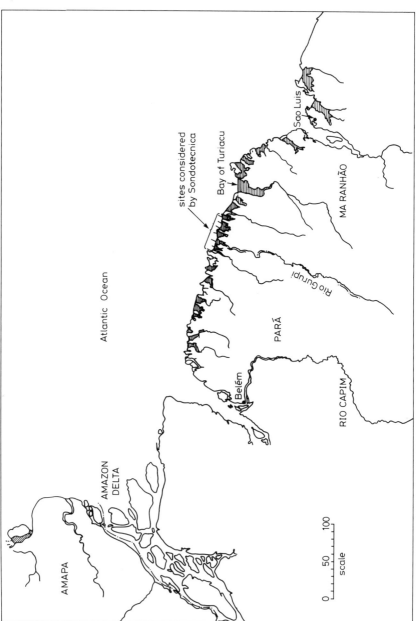

Fig. 12.19   Tidal resources along the north coast of Brazil

The Sondotecnica comparative figures are 3400 MW installed, giving an annual energy output of 9114 GWH. Both figures may be somewhat optimistic. From the Admiralty chart (No. 3958), it appears that most of the bays and inlets studied are shallow, with depths at low tides of less than 5 m and many large drying sandbanks. Shallowness helps reduce the cost of a barrage but at the expense of making it difficult to submerge the turbines satisfactorily. The large deposits of sediments along this coast could cause problems with siltation of any tidal basin.

There is an existing tidal barrage on the Rio Bacanga, near and to the west of Sao Luis, which comprises a set of three 12 m wide sluice openings, with radial gates, at one end of an embankment and road across the estuary. At the time of a visit by the author in 1982, the middle gate had been removed and the other gates were in poor condition (Fig. 12.20). The barrage was built to control water levels upstream and prevent flooding of agricultural land by excluding high tides. It lacks only turbines and a transmission link to convert it to a power-generating barrage. This would be much less costly than starting from scratch. Plans were announced in 1988 to carry out this conversion and install turbines totalling 36 MW. This would make this scheme the second largest in the world (Refs. 1981(9), 1987(3)).

Brazil has immense reserves of hydro-electric power available for development. Ref. 1981(10) shows that, in 1979, only 4% of the resources of the north region, and 45% of those in the north-east region, had been developed, producing together 17 TWh/year. On the other hand, only 2% of consumption was in the north, 70% being in the south. Thus transmission of power over long distances (and environmental aspects) will be a factor in the development of hydro-electric power, a factor which will count more against tidal power with its intermittent output.

Therefore, apart from the development of the Rio Bacanga scheme as a demonstration project, with costs reduced by the presence of an existing barrage, it seems that tidal power should not be developed further in Brazil in the foreseeable future.

12.6 Canada

In the north-east corner of Canada or, more precisely, Nova Scotia (Fig. 12.21), lies the Bay of Fundy which has the largest tides in the world. Fig. 12.22 shows a spring–neap cycle for Burncoat Head which has a mean tidal range of 11·28 m, the largest listed in Admiralty Tide Tables. What is also a feature of the Bay of Fundy is that the S2 constituent is relatively small. At Burncoat Head, for example, M2 is 5·64 m while S2 is only 0·83 m. Thus the mean spring range $(2 \times (M2 + S2))$ is 12·94 m while the mean neap range $(2 \times (M2 - S2))$ is exceptionally large at 9·62 m. Comparable figures for Clevedon in the Severn estuary are 11·06 m and 5·54 m. This points to a tidal power scheme having much less variation than normal over a spring–neap cycle. The disadvantage is that construction will remain difficult during neap tides, because currents will be fast and slack water periods will be short.

The Bay of Fundy has been the subject of extensive study over the last 15 years, and the results have been published in comprehensive reports and papers (Refs. 1977(1), 1979(5), 1980(7), 1985(1)). A large number of possible barrage sites has been studied and compared. The shortlisted sites are shown on Fig. 12.21 and their principal features have been used in the development of the parametric method described in the previous chapter. The preferred site is B9,

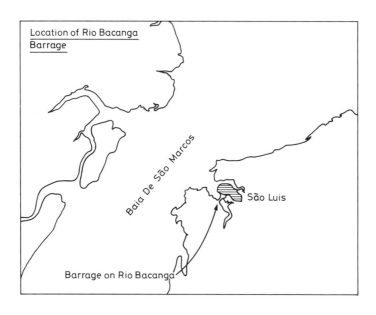

Fig. 12.20   Rio Bacanga Barrage

Fig. 12.20 *Continued*: Flow through gated section

close to Burncoat Head, and this has been uprated, following review studies, from 4028 MW to 4865 MW.

Although the Bay of Fundy is arguably the prime site for tidal power in the world, it suffers from remoteness from centres of demand. Consequently much effort has been spent in identifying methods of exporting the power. These tend towards a transmission link about 1000 km long to the south, to connect into the grid system of the State of New York in the USA.

On the east coast of the Bay of Fundy is the town of Annapolis Royal, located on the east shore of an inlet off the bay, formed by the Annapolis River. This is the site of the Annapolis tidal power station which was started in 1980 and completed in 1986. The station is the first in North America, and comprises a 7·6 m diameter Straflo turbine generator of 20 MW capacity, installed in a power house built in Hog's Island (Fig. 12.23). An existing barrage had been built to connect Hog's Island to the shore of the inlet to form a flow control barrier, releasing river flows and preventing flooding of agricultural land by high tides. The turbine is the prototype large Straflo machine, intended to demonstrate the concept both for tidal power and large run-of-river schemes. Since commissioning, high availability has been achieved (Ref. 1987(4)).

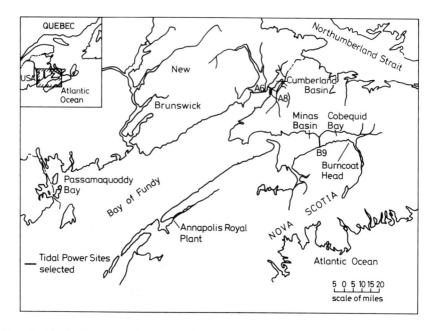

Fig. 12.21   Tidal power sites in the Bay of Fundy

Outside the Bay of Fundy, the north, Arctic, coast of Canada and the south coast of Baffin Island along Hudson Strait have sites with mean tidal ranges between 5 and 8·7 m, the largest being inside Ungava Bay (Fig. 12.24). On Baffin Island, Frobisher Bay has mean tidal ranges up to a maximum of 6·9 m at the town of Frobisher. The extreme remoteness and severe climate prevents these sites being considered seriously. If conditions were more favourable, then Leaf Basin, on the west side of Ungava Bay, with a mean tide range of 8·72 m would appear to be the first site to consider.

Surprisingly, the standard port under which the data for Hudson Strait and Ungava Bay are listed in Admiralty Tide Tables is Puerto Gallegos, one of the more interesting sites in Argentina and about 12 000 km to the south.

12.7 China

The south-east coast of China, opposite the north end of Taiwan (Fig. 12.25) has a mean tidal range of 4–4·5 m, peaking at 5·14 m at the island of Sandu Dao (Fig. 12.27). Although the coastline is very indented, like the west coast of South Korea, which is discussed later, most of the bays and inlets are shallow with extensive mudflats. This excludes many of the smaller inlets and limits the number of sites that can be considered for tidal power.

Further north, the mean tidal range along the coast is about 2 m, but, south

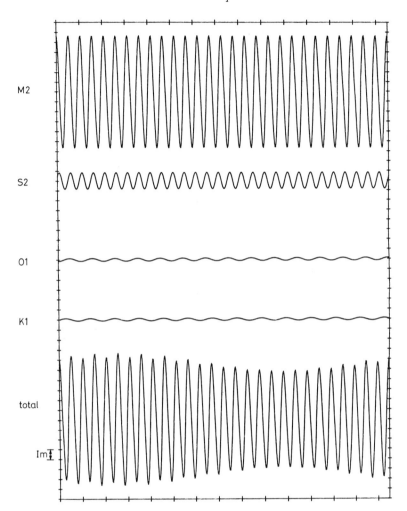

Fig. 12.22   Tides at Burncoat Head, Bay of Fundy

of Shanghai, the large estuary Hangshow Wan runs inland about 200 km. This allows the mean tidal range to increase to 5·8 m at Baitashan near the city of Hangchow. In this case, however, the topography is less favourable as regards the ratio of length of the barrage to the area of the enclosed basin.

Several small tidal power schemes have been built in China, but have not been reported in any detail in the Western technical press. The largest (Ref. 1981(8)) was built in the dry alongside Jianxia Creek on the Zhejiang coast. After the power house was complete, the creek was closed off with a dam about 15 m high and canals were dug to connect the power house to the basin and the sea. Some sluices were also provided. In 1981 the first of six 500 kW turbines

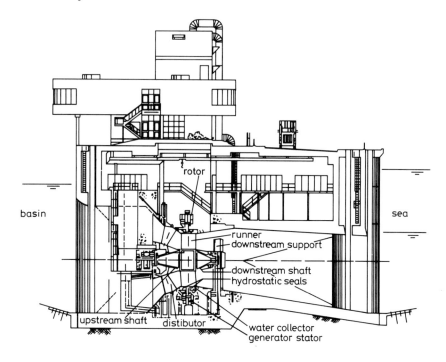

Fig. 12.23   Cross-section of power station at Annapolis Royal (Escher-Wyss, Zürich)

had been installed, so that the final capacity will be 3 MW. This scheme operates with two-way generation. Other reported schemes include two ebb generation plants, one of 40 kW commissioned in 1959 and since uprated to 200 kW, and one of 165 kW. Their locations cannot be readily identified on small scale maps.

Working from Admiralty charts Nos. 1761 and 1754, the sites shown in Figs. 12.26 and 12.27 have been assessed using the parametric method. The results are listed in Table 12.5.

The two sites considered for Damao Shan compare the effects of deepening and narrowing at the mouth of the estuary on the unit cost of electricity. The shallower site appears preferable, and should be easier to build. These results suggest that, for the sites with mean tidal range of about 5 m and reasonable depths of water, quite low unit costs of electricity (by western standards) could be achieved. If the same effect of low labour costs found in the South Korean work were to apply to China, then the unit costs listed above could perhaps be halved. The total resource represented above, not counting the output of sites located behind larger sites, is about 20 TWh/year.

These results have been based on the assumption that a barrage at each site would be operated as an ebb generations scheme. This would require an established electricity transmission grid, linking the barrage both to centres of demand and to either thermal plant or large hydro schemes which could be

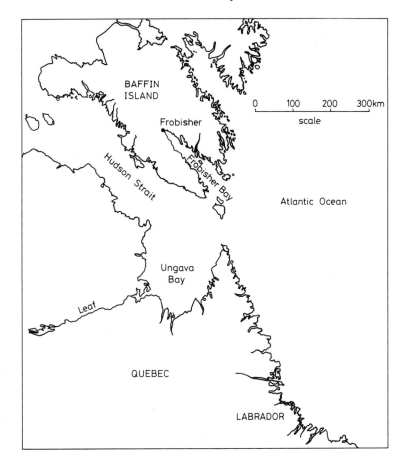

Fig. 12.24 North coast of Canada

cycled to absorb the power from the tidal scheme(s). In the absence of a suitable grid, the larger sites must remain undeveloped, while the smallest sites could perhaps be developed as two-way schemes, as has been done already in China at very small scale.

12.8 Eire

The west coast of Eire is heavily indented, but the tidal range is less than the target figure of 5 m considered necessary for tidal power to be economic. The estuary of the River Shannon, however, does have a large area and is narrow in places. At Ballyhoolahan Point (Fig. 12.28), a barrage would have the following characteristics, based on the parametric method of assessment:

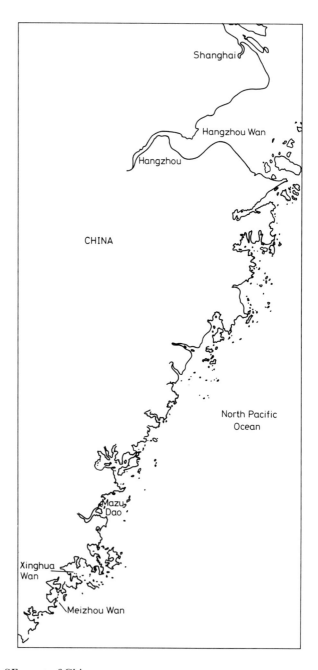

Fig. 12.25   SE coast of China

Mean tidal range	3·6 m
Area	200 km² approx.
Length of barrage	1600 m
No. of turbines × dia.	45 × 7·5 m
Installed capacity	400 MW
Annual energy output	820 GWh
Unit cost of energy	5p/kWh

Although representing a sizeable national resource, the unit cost of electricity is not low enough to be attractive.

12.9 France

The 240 MW tidal power station on the estuary of the river Rance near St. Malo (Fig. 12.29) is the largest such scheme in the world, and credit must be given to the French engineers who designed and built this imaginative project. Although of considerable size, the Rance barrage was built at a time of

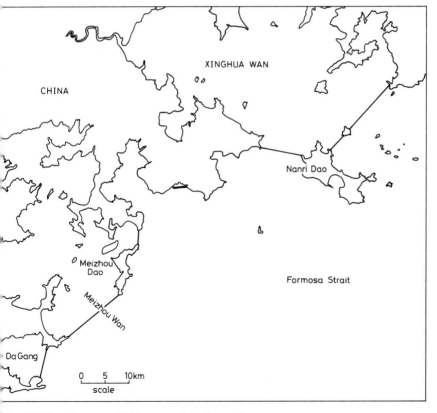

Fig. 12.26 SE coast of China: possible sites of barrages

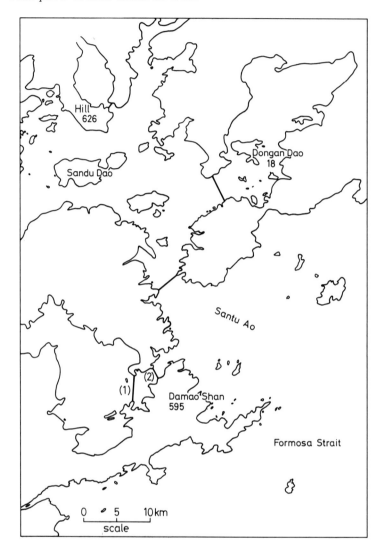

Fig. 12.27   SE coast of China: possible sites of barrages

uncertainty in France concerning the future sources of power, and was therefore intended partly to be a prototype for a much larger scheme to seaward. This is the Iles de Chaussee scheme, of which several variations have been suggested, including two-basin schemes designed to smooth the power output somewhat and make it easier to absorb. These are discussed in Chapter 2.

Table 12.5 Sites on coast of China

Site	Area (km²)	Max. depth (m)	Length (m)	Mean range (m)	Turbines (no. × dia.)	Annual energy (TWh)	Unit cost (p/kWh)
Da Gang	52	12	4550	4·0	52 × 4	0·36	5·5
Meizhou Wan	550	24	13 200	4·0	250 × 6	3·83	4·7
Xinghua Wan	930	30	20 500	4·1	185 × 9	6·7	5·0
Sites in Fig. 12.27							
Damao Sahan (1)	200	24	3550	4·8	100 × 6	2·05	3·7
Damao Shan (2)	210	45	1400	4·8	46 × 9	2·15	4·3
Hill 626	19	10	620	5·1	20 × 4	0·20	2·8
Dongan Dao	210	21	3900	5·1	100 × 6	2·26	3·2
Santu Ao	680	35	3000	4·8	150 × 9	3·7	2·8

12.10 India

There are two inlets on the west coast of India which have tidal ranges large enough to be of interest for tidal power. These are the Gulf of Kachchh and the Gulf of Cambay or Khambhat (Fig. 12.30). The Gulf of Kachchh has been the subject of investigations and studies carried out by the Tidal Power Cell of the Central Electricity Authority (CEA) who have published some results in Ref. 1986(8). The site selected is reasonably narrow in relation to the area of the enclosed basin and would be deep enough to accommodate 6 m diameter turbines (Fig. 12.31). The main features are summarised in Table 12.6.

The Gulf of Cambay has a larger tidal range than the Gulf of Kachchh. At the port of Bhavnagar the mean range is 6·28 m with a pronounced semi-diurnal inequality (Fig. 12.32). This range is above the minimum range of 5 m considered necessary for economic development. The area enclosed and the length and depth of the site are much greater than those at the Gulf of Kachchh, and so this scheme would be a major undertaking.

Both sites are in estuaries which contain extensive areas of sandbanks and mudflats. This raises a question regarding siltation of the enclosed basin once a barrage were operating. In addition, the rivers flowing into the basins appear to be able to transport large quantities of sediments during floods. For example, the maximum observed floods in the Sabarmati and Mahi Sagar rivers entering the Gulf of Cambay have been 15 000 and 40 000 m³/s, respectively. These are an order of magnitude larger than the river Severn.

The presence of extensive sandbanks suggests that any barrage would have to be designed to be built on soft foundations rather than rock. In this event, extensive measures to prevent scour would be needed; this could take the form of stone-filled mattresses on geotextile filters.

Two sites in the Gulf of Cambay have been assessed using the parametric method (Fig. 12.31). They were looked at because these were alternatives being considered in 1988 for a road crossing of the estuary. The results of the assessments are shown in Table 12.6.

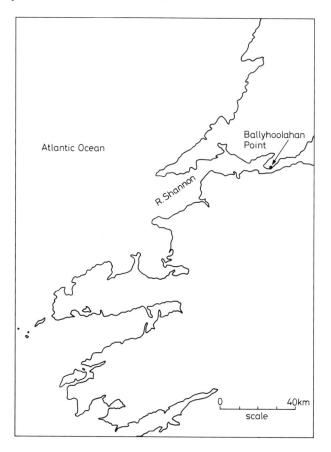

Fig. 12.28   West coast of Eire and R. Shannon

The tentative conclusion is that the Gulf of Cambay represents a very large tidal power source, with potentially low unit costs of energy. Bearing in mind that the unit costs are based on UK construction costs, there will be scope for these to be reduced by employing local labour, at least on the civil engineering structures. Studies of Garolim Bay in South Korea (Ref. 1981(16)) showed that the unit cost of electricity could be about half that of an equivalent scheme in the UK.

The Gulf of Kachchh does not appear to be as attractive as the Gulf of Cambay, at least as regards the unit cost of electricity. However, with an installed capacity about one twentieth of the Cambay schemes, this would be a much more manageable investment and risk, and further studies and design are planned.

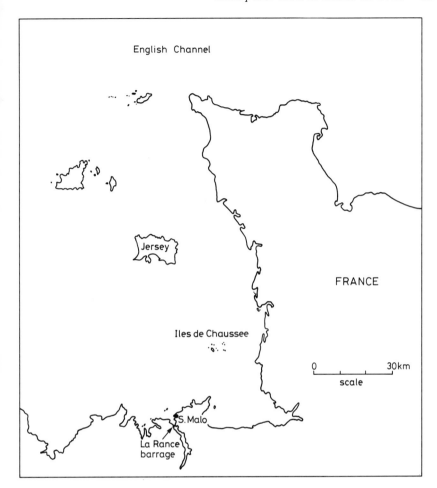

Fig. 12.29 French tidal power sites

12.11 Northern Ireland

On the east coast of Northern Ireland, south of Belfast, lies Strangford Lough (Fig. 12.33). The possibility that this large tidal estuary, which has an area of about 140 km^2 and a very narrow mouth in proportion to its area, could be developed for tidal power has been studied by the Northern Ireland Economic Council (Ref. 1981(3)). The results, which have been included in the data on which the parametric method of assessment is based (Chapter 11), were disappointing because the unit cost of tidal energy was estimated at 7p/kWh. This high figure results from the mean tidal range being only 3·1 m.

The large areas of intertidal flats in Strangford Lough are an important feeding area for wading birds, while the sediment regime appears to be more

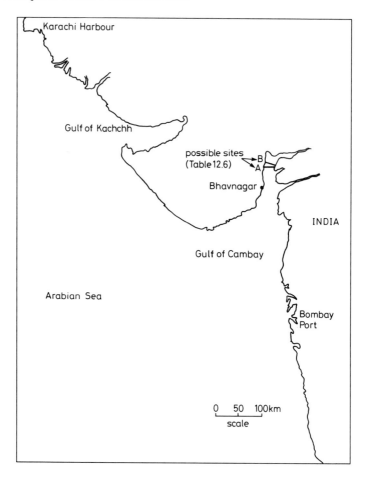

Fig. 12.30   West coast of India: Gulf of Kachchh and Gulf of Cambay

stable than, say, that in the Severn estuary. Thus the loss of intertidal feeding areas is of concern to environmentalists, while no offset by means of a more stable and productive sediment regime can be offered.

12.12 South Korea

The west coast of South Korea has mean tidal ranges reaching 6 m. The coast is also highly indented and therefore appears to offer various possibilities for tidal power development (Fig. 12.34). However, many of the bays are shallow, with extensive intertidal flats which are being reclaimed in order to increase the area of land available for cultivation.

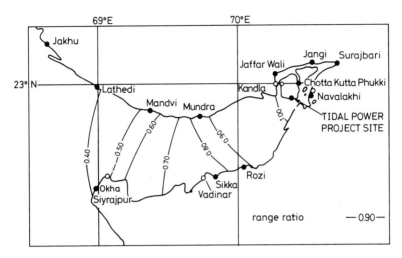

Fig. 12.31 Co-tidal chart for Gulf of Kachchh

Two possible sites have been considered. The first, Garolim Bay, has been the subject of detailed feasibility studies carried out on behalf of, and by, the Korean Ocean Research and Development Institute (KORDI). Initial studies involved the French company Sogreah (Ref. 1981(4)), and the results were reviewed and revised in a second study by KORDI assisted by a consortium of British companies who had been involved in the studies of the Severn barrage (Ref. 1986(13)). This review included the optimisation of the operation of the barrage as an ebb generation scheme, with tides having a pronounced semi-diurnal inequality. Fig. 12.35 shows typical results.

The second site, in the Gulf of Asam, has apparently not been studied in any detail. It has a mean tidal range of 6·06 m, larger than that at Garolim Bay, and therefore should be more economic. The main features of the two sites are summarised in Table 12.7.

These results suggest that the Gulf of Asam is the better site of the two and, if the unit cost of electricity could be reduced by the same factor as Garolim Bay, this should be an attractive location.

Table 12.6 Sites in Gulf of Cambay and Kachchh

Site	Area (km²)	Max. depth (m)	Length (m)	Mean range (m)	Turbines (no. × dia.)	Annual energy (TWh)	Unit cost (p/kWh)
Cambay A	740	18	24 000	6·1	400 × 6	11·5	2·5
Cambay B	1055	22	25 000	6·1	570 × 6	16·4	2·5
Kachchh	50?	18	2000	4·8	24 × 6	0·48	5

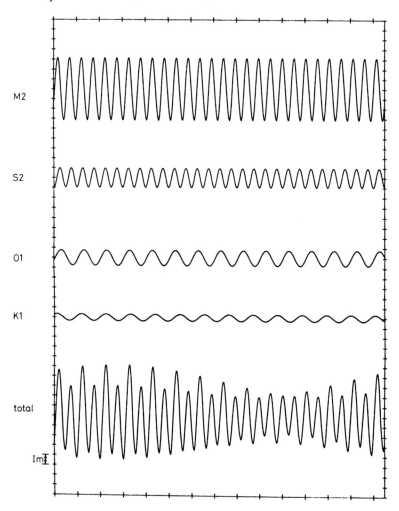

Fig. 12.32   Tides at Gulf of Cambay: Bhavnagar

12.13 United States of America

The west coast of the United States is of little interest as regards tidal power. The mean tidal range is generally between 1 and 2 m and is diurnal. At the north end of the Gulf of California, the mean range increases, reaching a maximum at the mouth of the Colorado River of about 5 m (Fig. 12.36). However, the bay is shaped such that any barrage would have to be very long in proportion to the enclosed area, while the Colorado river mouth contains large areas of sandbanks.

On the east coast, the only area of interest is the southern part of the Bay of Fundy. Passamaquoddy Bay is a complex area of inlets and islands where the

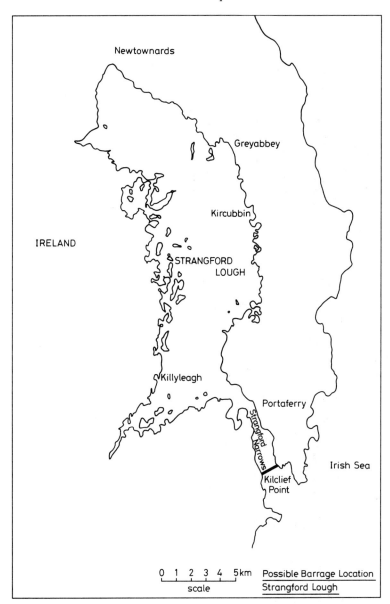

Fig. 12.33 Strangford Lough

mean tidal range is about 5·4 m (Fig. 12.21). As is to be expected, this is much less than the ranges at the top of the Bay of Fundy and, based on the parametric method of assessment, is unlikely to result in economic power generation unless there is a site with a large area enclosed in relation to the length of the barrage. Nevertheless, plans were made and work was started on a 110 MW capacity

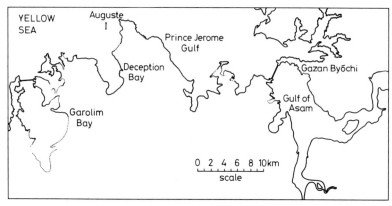

Fig. 12.34   West coast of S. Korea: Gulf of Asam, Garolim Bay

two-basin scheme in 1935, during the depression (Refs. 1941(1), 1978(5)). This was stopped in 1936, after work to the value of $6 million had been completed, when comparisons were made between the unit cost of energy from the tidal scheme and that from normal hydro-electric schemes in the State of Maine. The latter were found to be a better investment.

12.14 United Kingdom

Apparently not well known, but perhaps not surprising, is that the United Kingdom had a tidal power scheme which generated electricity and smoothed the output by a form of pumped storage operating in 1930 (Ref. 1930(1)). The

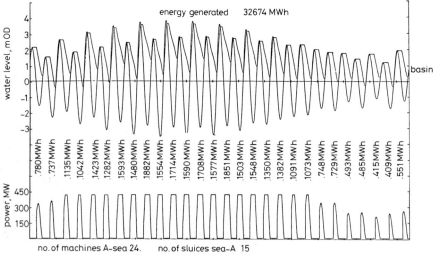

Fig. 12.35   Operation of Garolim Bay, tidal power scheme

Table 12.7 Sites in S. Korea

Site	Area (km²)	Max. depth (m)	Length (m)	Mean range (m)	Turbines (no. × dia.)	Annual energy (TWh)	Unit cost (p/kWh)
Garolim Bay	100	28	1850	4·8	24 × 8	0·893	4·5*
Gulf of Asam	130	24	2350	6·06	72 × 6	2·05	3·1*

* Based on parametric method. Actual estimated cost for Garolim Bay is 2·1p/kWh, largely due to low labour costs.

inventor was a Mr P. Shishkoff, a Russian immigrant, and it was installed in one of the Avonmouth docks, syphoning water over the dock wall. The water turbine was by Boving & Co., with a maximum output of about 350 kW. The arrangement of the plant is shown in Fig. 12.37. It had a continuous output of 16 kW. Losses and leaks proved unsatisfactory.

The west coast of England and Wales offers a wide range of tidal power sites. Following the first oil crisis in 1974, and following wide ranging discussions concerning the prospects for tidal power (e.g. Ref. 1977(2)), much effort has been expended since 1978 on studies to define the size of the resource, identifying the most favourable sites and assess their economic and environmental feasibility. Possibilities range from a 12 000 MW barrage in the outer part of the Severn Estuary (dropped in favour of an 8000 MW barrage further landwards) to barrages with capacities of 5 MW or less.

The results of these studies have been published (Refs. 1980(1), 1981(2), 1986(1)) and so only a brief summary is given here. The sites divide into two categories: large (about 500 MW or above) and small. The large estuaries, namely the Severn, Solway Firth, Morecambe Bay, Dee and Mersey and, on the east coast, the Humber, Wash and Thames, have each been the subject of preliminary studies. The Severn and the Mersey have been and are being studied in great detail as they are the estuaries most likely to be economic. The results have been used to develop the parametric method of assessment described in the previous chapter. Fig. 12.38 summarises the results.

The small estuaries, including some along the south coast of England, have been assessed and ranked in terms of their likely output and the unit cost of that output, using the parametric method, and their energy output and cost are included in Fig. 12.38. The locations of some of the best are shown in Fig. 12.39. None of these sites is likely to be an economic investment in its own right, but the benefits of increased depths of water in the enclosed basin, protection against flooding by exceptionally high tides, and the opportunity for development of the estuary could justify the investment needed to make up the difference between the return from sales of electricity and the cost of financing the construction and operation of the barrage.

Although each possible site would be of great interest locally, what is clear from these results is that the three best large estuaries, namely the Severn, Solway Firth and Morecambe Bay, dominate the tidal power scene in the UK.

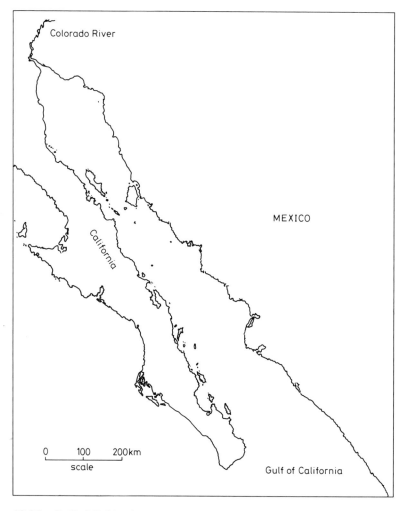

Fig. 12.36   Gulf of California

They have the added advantage that high water in the Solway Firth and Morecambe Bay is about five hours later than in the Severn, so that power from the last would be produced between periods of generation at the two northern estuaries.

12.15 USSR

There are two parts of the Russian coast which are of interest for tidal power. On the east coast, the Sea of Okhotsk is a large embayment about 1500 km long. At its northern extreme is the Gulf of Penzhinskaya (Fig. 12.40), where the

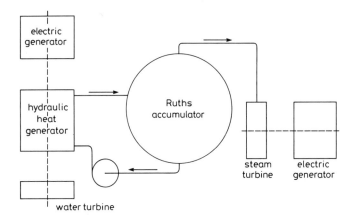

Fig. 12.37 Diagram of plant arrangement of 1930 Shishkoff scheme

spring tidal range exceeds 10 m. Fig. 12.41 shows a 14.5 day cycle for Anstronomicheski Point on the south shore near the top of the Gulf. However, as can be seen, the tides are diurnal with the largest constituent being K1 with an amplitude of 2·52 m. At the western end of the Sea of Okhotsk is a complex area of islands and inlets (Shantarskiye Ostrova, Fig. 12.42) where the tides are semi-diurnal with a pronounced diurnal inequality (Fig. 12.43). The mean tidal range is only about 3·6 m, so this area would not meet the criterion of 5 m mean range developed earlier. However, in a recent article (Ref. 1986(3)) Bernstein,

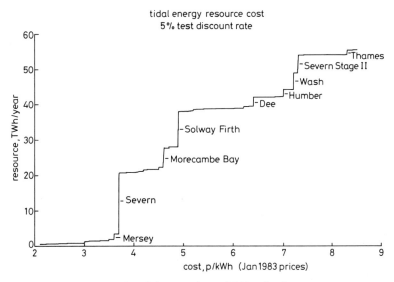

Fig. 12.38 Cumulative UK tidal energy/cost (1983 prices)

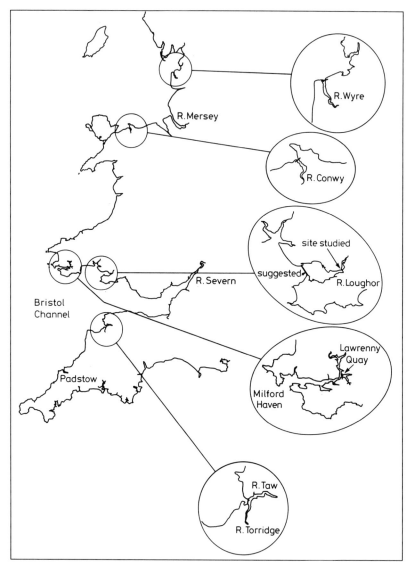

Fig. 12.39   Locations of some small sites, worthy of further study

one of the fathers of tidal power, has suggested various schemes for these locations, all of which would be very large.

The parametric method of estimating the size and output of a tidal power scheme which has been applied in this chapter was developed from results of studies of barrage sites where the tides are semi-diurnal. Thus it cannot be applied directly to a site with diurnal tides. However, estimates can be made, based on the following assumptions:

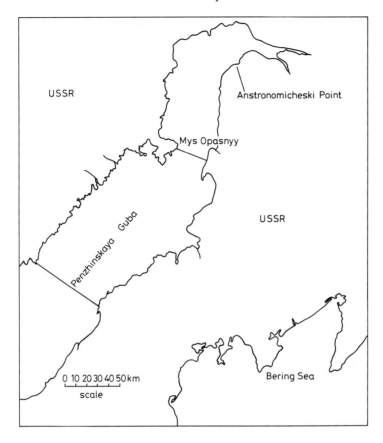

Fig. 12.40 Possible large barrage sites: north part of Sea of Okhotsk

- The basic rules for ebb generation still apply; namely that the basin should be filled to as high a level as possible, that there should be a slack period after high water while the sea ebbs enough for the turbines to start at reasonably high efficiency, and that enough turbines are installed to enable the basin to be drawn down to about mean sea level at the end of the generating period.
- Following from this method of operation, power will be generated for about eleven hours each day during spring tides and six hours during neaps, i.e. for the same tidal time per day as a normal semi-diurnal scheme, but the rate of releasing water will be half that for the equivalent semi-diurnal scheme and so the power will be halved
- The number of turbines will be half that for the equivalent semi-diurnal scheme.

On this basis, the parametric method has been applied to three possible sites, as listed in Table 12.8.

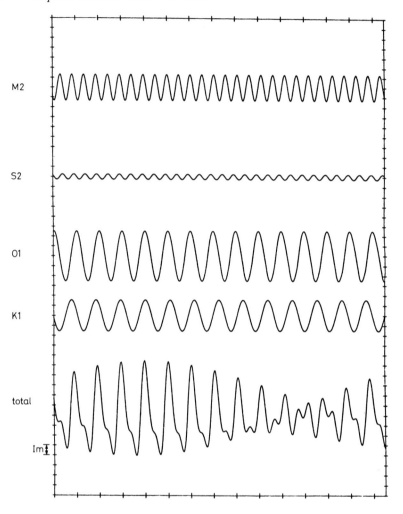

Fig. 12.41   Tides at Myś Astronomicheski

These results are subject to more than the usual uncertainty, partly because the Admiralty chart for this area lacks detail.

Of the three sites listed in Table 12.8, the mean tidal range for the first is too small for it to be feasible, as indicated by the unit cost of electricity. The second site has been evaluated for a mean tide range of 4·75 m, this being the value quoted by Bernstein, and the cost of electricity is much lower but unlikely to be competitive with large river hydro-electric projects. Admiralty Tide Tables show a mean tide $(2 \times M2)$ of 3·08 m at the point on the north end of the east bank, so the tide to seaward of the barrage may be reduced by the barrage itself. The third site, with a diurnal tide averaging about 5 m, appears more suitable. In his review article, Bernstein prefers two-way generation and suggests an

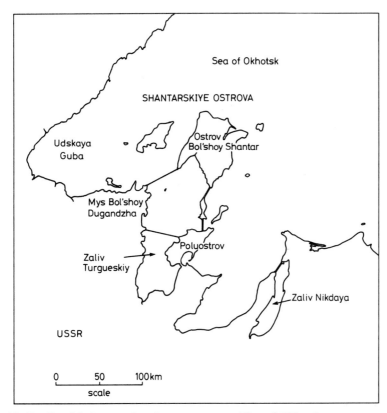

Fig. 12.42 Possible barrage locations: west part of Sea of Okhotsk

installation for the third site which includes 568 10 m diameter turbines and 920 7·5 m diameter machines, with a total energy output of 79 TWh per year from 20·4 GW. The ratio of energy output to installed capacity, nearly 4:1, is much higher than the normal figure of about 2:1. The unit cost of electricity is uncertain because of the effect of the diurnal tide. If the tide was semi-diurnal, the installed capacity would be doubled and the unit cost would be about

Table 12.8 Sites in the USSR

Site	Area (km²)	Max. depth (m)	Length (m)	Mean range (m)	Turbines (no. × dia.)	Annual energy (TWh)	Unit cost (p/kWh)
Zaliv Nikolaya	670	30?	8500	3·08	100 × 9	2·43	6
Zaliv Turgurskiy	1400	30	26 000	4·74	200 × 9	12	4
Mys Opasnyy	9200	65?	36 000	5 (D)	1090 × 9	54	4?

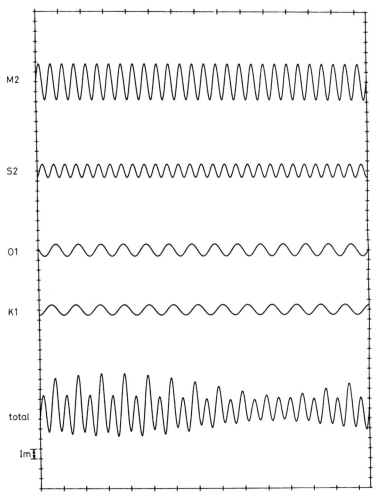

M2

S2

O1

K1

total

Im

Fig. 12.43   Tides at Guba Lebjazh'ya

3·5p/kWh. Even with 1000 turbines there would be problems in accommo-
dating this number in the space available. However, the depth of water at the
site is enough to allow turbines to be double-banked and/or have sluices above
or below the turbines.

Table 12.9   Site on the White Sea

Site	Area (km²)	Max. depth (m)	Length (m)	Mean range (m)	Turbines (no. × dia.)	Annual energy (TWh)	Unit cost (p/kWh)
Mezenskiy Zalev	2300	40?	90 000	5	520 × 9	25·4	6

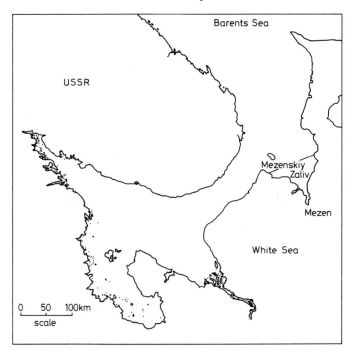

Fig. 12.44   Plan of White Sea, showing location of Gulf of Mezen

One problem with these sites is the severe winter conditions. Thick ice can develop and would cause large loads on the structures.

The only other part of the coast of the USSR with any reasonable prospects for tidal power is the White Sea on the north coast (Fig. 12.44). At the mouth, the mean (semi-diurnal) tidal range is about 3 m. This range increases to 5·48 m in the south corner of the Gulf of Mezen, and it is near Mezen that the 400 kW Kislaya Guba tidal plant was prefabricated and floated into position in 1969. At the entrance to the south western part of the White Sea the mean range is about 4·5 m and decreases rapidly to about 1 m further southwards. Thus the only area offering reasonable prospects is the bay labelled Mezenskiy Zaliv. Bernstein has proposed a barrage 90 km long enclosing much of this bay. Applying the parametric method to this site, the results given in Table 12.9 are obtained.

Again, the annual energy is half that suggested by Bernstein. As a result, the unit cost of energy is not likely to be economic and this site also suffers from a severe climate.

References

1923(1) MEIK, C.S., ADDENBROOKE, G.L. and TWINBERROW, J.W.: 'Utilisation of waterways for the production of power; its consequences and applications, with special reference to the river Severn'. XIIIth International Congress of Navigation, 1st Section, London

1930(1) ANON.: 'A tidal power demonstration plant', *Power Engineer*, Nov. pp 427–431

1933(1) 'Report of the Severn Barrage Committee', (HMSO)

1941(1) FEDERAL POWER COMMISSION, Washington: 'Passamaquoddy Tidal Power Project, Maine', 1941

1945(1) VAUGHAN-LEE, A.G., HALCROW, SIR W. and DONKIN, S.B. For Ministry of Fuel and Power: 'Report on the Severn Barrage scheme'. (HMSO)

1948(1) RICHARDS, B.D.: 'Tidal power: Its development and utilisation', *Proc. ICE*, Paper 5656, pp. 104–131 and discussion pp. 131–149

1949(1) HEADLAND, H.: 'Tidal power and the Severn Barrage', *Proc. IEE* 96, Part 2, (51)

1962(1) FRAENKEL, P.M. and TRIGGS, R.L.: 'Special features of the civil engineering works at Aberthaw Power Station'. *Proc. ICE*, Paper 6599

1963(1) LEWIS, J.G.: 'The tidal power resources of the Kimberleys'. *J. Inst. Engs., Australia*, 35, (12), pp. 333–345

1963(1) HYDRAULICS RESEARCH STATION, Wallingford: 'Port Talbot, South Wales: Storm and long period wave data'. Report EX 221, August

1964(1) DRONKERS, J.J.: 'Tidal computations in rivers and coastal waters'. (North-Holland)

1965(1) ZANVELD, A.: 'The delta plan schemes'. Acier Stahl Steel, No 12, December

1965(2) BERNSTEIN, L: 'Tidal energy for electric power plants'. Israel Program for Scientific Translations, Jerusalem, pp. 208–344

1965(3) FORD, S.E.H., and ELLIOT, S.M.: 'Investigation and design of the Plover Cove water scheme'. *Proc. ICE*, Paper 6881, Oct.

1966(1) GIBRAT, R.: 'L'energie des marées'. (Presses Universitaires de France, Paris)

1968(1) HEAPS, N.S.: 'Estimated effects of a barrage on tides in the Bristol Channel'. *Proc. ICE*, 40, pp. 495–509

1970(1) 'Dredging today'. Vereniging 'Centrale Baggerbedrijf' The Hague, pp. 70–75

1972(1) FERGUSON, H.A.: 'The Netherlands Delta project: Problems and lessons', *Proc. ICE*, Paper 7449, March

1973(1) HOLLOWAY, B.G.R. *et al.*: 'Passage gates for Seaforth Dock, Liverpool', *Proc. ICE*, Paper 7599, May

1973(2) AGAR, M., and IRWIN-CHILDS, F.: 'Seaforth Dock, Liverpool: Planning and design', *Proc. ICE*, Paper 7600, May

1973(3) COLE, P.G., *et al.*: 'Seaforth Dock, Liverpool: Construction', *Proc. ICE*, Paper 7601, May

1974(1) BUILDING RESEARCH ESTABLISHMENT for DoE: 'A survey of the locations, disposal and prospective uses of the major industrial by-products & waste materials'. Feb.

1976(1) MAUNSELL & PARTNERS in association with Binnie & Partners and Mertz & McLellann & Partners for State Energy Commission of Western Australia: 'Kimberley Tidal Power'. Jan.

1976(2) RAMSAR: 'Convention on wetlands of international importance especially as wildfowl habitat', (HMSO, May)

1976(3) LILWALL, R.C.: 'Seismicity and seismic hazard in Britain'. (Institute of Geological Sciences)

1976(4) BINNIE & PARTNERS: for Central Water Planning Unit of DoE: 'The Wash water

storage scheme: Feasibility study: Vol. 2, Surveys'.
1976(5): *Ibid.* 'Vol. 10, Trial Banks'.
1977(1) BAY OF FUNDY REVIEW BOARD & MANAGEMENT COMMITTEE: 'Reassessment of Fundy tidal power'. Nov.
1977(2) Informal discussion: 'The prospects of tidal power'. *Proc. ICE*, Part 1, (62), Nov., pp. 701–705
1977(3) ANON.: 'On the Eider', *Consulting Engineer*, Oct. 77 pp. 24–25
1977(4) HOLLANDOCHE BESTON GROEP NV.: 'The history of caisson construction within the HBG from 1902 to 1977'
1978(1) NEDECO for Dept. of Energy. 'Severn Tidal Barrage Scheme: Study on the shortening of the construction time'. Report No STP 1, April
1978(2) NORTHUMBRIAN WATER AUTHORITY: 'Kielder Water: Comprehensive plan for the development of recreation and tourism. Nov.
1978(3) FONG S.W., and HEAPS, N.S.: 'Note on quarter-wave resonance in the Bristol Channel'. Institute of Oceanographic Sciences Report No. 63
1978(4) THOMASSON, W.L., and REYNOLDS, H.L.: 'Multipurpose development of the Arkansas River Project includes hydroelectric power stations of unusual design'. Allis Chalmers
1978(5) CHARLIER, R.H.: 'Tidal power plants: Sites, history and geographical distribution', Proc. International Symposium on Wave and Tidal Energy, Canterbury, England, Sept. BHRA
1979(1) BINNIE & PARTNERS for Dept. of Energy: 'Severn tidal power — Embankments'. Report No STP 12. June
1979(2) ENGINEERING AND POWER DEVELOPMENT CONSULTANTS for Dept. of Energy: 'Severn tidal power – Selection of gates. Initial Review'. Report No STP 19, May
1979(3) BINNIE & PARTNERS for Dept. of Energy: 'The closure of an ebb generation barrage'. Report No STP 36, May
1979(4) DELTAMARINE CONSULTANTS BV.: 'Comparison of caisson placing in high current tidal areas'. Report No 950034/04, Sept.
1979(5) FUNKE, E.R., PLOEG, J., and CROOKSHANK, N.L.: 'Modelling requirements & techniques for tidal power development in the Bay of Fundy'. National Research Council, Canada, Division of Mechanical Engineering, Ottawa, March
1979(6) COCHRANE, G.H., *et al.*: 'Dubai Dry Dock: design and construction', *Proc. ICE*, Paper 8202, Feb.
1979(7) UNIVERSITY OF SALFORD (Dept of Civil Engineering) for Dept. of Energy: 'Tidal energy survey of the Severn Estuary and Bristol Channel (single basin, single effect), & supplementary reports'. Reports STP 6, 7, 8, Jan–Sept
1979(8) LLEWELLYN, T.J., and MURRAY, W.T.: 'Harbour works at Brighton marina: construction', *Proc. ICE*, Paper 8242, May
1980(1) BINNIE & PARTNERS for Dept. of Energy: 'Preliminary survey of tidal energy of UK estuaries'. Report STP 102, May
1980(2) INSTITUTE FOR MARINE ENVIRONMENTAL RESEARCH for Dept. of Energy: 'Severn Tidal Power – Predicted effects of proposed tidal power schemes on the Severn Estuary Ecosystem: Vol 1: Water quality'; 'Vol 2: Ecosystem effects'. Report Nos STP 37 & 38, Nov.
1980(3) HYDRAULICS RESEARCH STATION, Wallingford for Dept. of Energy: 'Severn tidal power: 2-D Water movement model studies'. Report EX 985
1980(4) BINNIE & PARTNERS for Dept. of Energy: 'The development of a one-dimensional model of the estuary'. July
1980(5) BINNIE & PARTNERS for Dept. of Energy: 'One-dimensional model studies of the Severn Estuary'. Oct. Also Supplementary Report, 1981
1980(6) WATER RESEARCH CENTRE for Dept. of Energy: 'Predictions of near-field dispersions at two sewage outfalls into the Severn with a tidal power barrage'
1980(7) DOUMA, A.: 'The Annapolis Tidal Power Project'. Thermal Engineering Dept., Nova Scotia Power Corporation, August
1980(8) FERNS, P.N., for Dept. of Energy: 'Research on wading birds and Shelduck in the Severn Estuary'. Report STP No 47, July
1980(9) SIR ROBERT McALPINE & SONS LTD. for Dept. of Energy: Further report on caissons and other civil engineering work'. Report STP 55. Sept.
1980(10) TAYWOOD ENGINEERING for Dept. of Energy: 'Severn tidal power: Caisson studies Stage 2: Report on the design, construction & placing of caissons'. Report STP 80, Nov.

1980(11) ATKINS PLANNING for Dept. of Energy: 'A study of recreational benefits of the proposed (Severn) barrage'. Report STP 96. Dec.

1980(12) INSTITUTE OF OCEANOGRAPHIC SCIENCES for Dept. of Energy: 'Tidal barrage calculations for the Bristol Channel: (i), (ii) & (iii)'. Report STP 15, 16 & 17, Mar.–Apr.

1980(13) SONDOTECNICA for Electrobras: 'Projecto conceitual de usina maremotriz'. Dec.

1980(14) SHAW, T.L.: 'Two-basin tidal barrages; incorporating a broad review of their electrical benefits in networks based on large fossil & nuclear plant'.

1980(15) BICKLEY, D.T. (University of Bristol) for Dept. of Energy: 'Energy studies of two-basin schemes – Severn barrage'. Report No. STP 26, July

1980(16) LEWIN, J. (J. Lewin & Partners): Personal communication

1980(17) RENDEL, PALMER & TRITTON for Dept. of Energy: 'STP – navigation aspects'. Report No. STP 79, Dec.

1980(18) RILEY, D.J. and SYMONDS, J.D.: 'Summary of fisheries interests of the Bristol channel east of Worms Head – Morte Point and the effects of alternative barrages at positions 2 and 5'. MAFF, Lowestoft

1981(1) SEVERN BARRAGE COMMITTEE: 'Tidal power from the Severn Estuary (2 Vols.)'. HMSO Energy Paper No 46

1981(2) WISHART, S.J.: 'A preliminary survey of tidal energy from 5 UK estuaries'. Proc. 2nd International Symposium on Wave and Tidal Energy, Cambridge, UK, BHRA, Sept.

1981(3) NORTHERN IRELAND ECONOMIC COUNCIL: 'Strangford Lough tidal energy', August

1981(4) SOGREAH for Korean Electric Company: 'Garolim tidal power plant feasibility studies'. Nov.

1981(5) WELSH WATER AUTHORITY for Dept. of Energy: 'The effects of a tidal barrage on migratory fish'. Report No. STP 30, Jan.

1981(6) UNIVERSITY OF BRISTOL for Dept. of Energy: 'Preliminary model tests of sluice caissons'. April

1981(7) NATURE CONSERVANCY COUNCIL: 'Severn tidal power: Nature conservation'

1981(8) ANON.: 'China looks to tidal power', *Elect. Rev.* 208, (3), 23 Jan.

1981(9) ELECTROBRAS: 'Usina maremotriz – estuario do Bacanga; Projecto Conceitual'

1981(10) MINISTRY OF MINES AND ENERGY (Federal Republic of Brazil): 'Brazilian energy model'. May

1981(11) PARKER, W.R. and KIRBY, R. for Dept. of Energy: 'The behaviour of cohesive sediment'. Institute of Oceanographic Sciences, Report No. STP 33

1981(12) THICKE, R.H.: 'Practical solutions for draft tube instability', *Water Power & Dam Construction*, Feb. pp. 31–37

1981(13) HYDRAULICS RESEARCH STATION, Wallingford for Dept. of Energy: 'Hindcasting extreme waves'. Report EX 978, March

1981(14) WHEELER, S.J.: 'Optimisation of tidal power schemes'. Proc. Second International Symposium on Wave and Tidal Energy, Cambridge, UK, BHRA. Sept.

1981(15) STOKES, C.J. and STREET, D.J.: 'Turbine caissons for the Severn Barrage'. Proc. Second International Symposium on Wave and Tidal Energy, Cambridge, UK, BHRA, Sept.

1981(16) EVANS, J.J., & PUGH, D.T.: 'An analysis of sea-level data at Flat Holm, Severn Estuary. IOS Report No 129, 1981

1981(17) KEILLER, D.C. and THOMPSON, G.: 'One-dimensional modelling of tidal power schemes'. Proc. Second International Symposium on Wave and Tidal Energy, Cambridge, UK, BHRA, Sept.

1981(18) PROCTOR, R.: Mathematical modelling of tidal power schemes in the Bristol Channel'. Proc. Second International Symposium on Wave and Tidal Energy, Cambridge, UK, BHRA, Sept.

1981(19) DERBELIUS, C.A. and WITHERELL, R.G.: 'A preliminary assessment of Cook Inlet tidal power'. Proc. Second International Symposium on Wave and Tidal Energy, Cambridge, UK, BHRA, Sept.

1982(1) BANAL, M.: 'Tidal energy in 1982'. *La Houille Blanche*, (5/6)

1982(2) PARKER, W.R. and KIRBY, R.: 'Sources and transport patterns of sediment in the inner Bristol Channel and Severn Estuary: Severn Barrage'. (Thomas Telford)

1983(1) MAKELA, G.A.: 'Float-in powerhouses', *ASCE J. Energy Eng*, 109, (2), June

1983(2) ANON.: 'Louisiana's first prefabricated low head hydroelectric station', *International Power Generation*, May

1983(3) RATCLIFFE, A.T.: 'The basis and essentials of marine corrosion in steel structures', *Proc. ICE*, Part 1, 74 pp. 889–907

1983(4) WILSON, E.M.: 'The case for the Severn barrage. Evidence submitted to the Sizewell B enquiry'. June

1984(1) BINNIE & PARTNERS for Dept. of Energy: 'Preliminary study of small scale tidal energy; Phases 1 to 3'

1984(2) ANON.: 'Prefab power', *Civil Engineering/ASCE*, July, pp. 42–43

1984(3) HILLAIRET, P., ALLEGRE, J. and MEGNINT, L.: 'La Rance tidal power scheme'. Notes of lecture presented at IEE, London, Oct.

1984(4) BOUCLY, F. and FUSTER, S.: 'Tidal power; Present French designs', *La Houille Blanche*, (8) pp. 597–606

1984(5) ERBISTE, P.C.: 'Estimating gate weights'. *Water Power & Dam Construction*, May

1985(1) DABORN, G.R.: 'Environmental implications of the Fundy Bay tidal power development', *Water Power & Dam Construction*, April, pp. 15–19

1986(1) BAKER, A.C., and WISHART, S.J.: 'Tidal power from small estuaries'. Proc. 3rd International Symposium on Wave, Tidal, OTEC and Small Scale Hydro Energy, Brighton, UK, BHRA. May

1986(2) DELORY, R.P.: 'The Annapolis tidal generating station'. Proc. 3rd International Symposium on Wave, Tidal, OTEC and Small Hydro Energy, Brighton, UK, BHRA. May

1986(3) BERNSTEIN, L.: 'Tidal power engineering in the USSR', *Water Power & Dam Construction*, March, pp. 37–41

1986(4) MOTOKI, H. *et al.*: '65 MW bulb turbine for Japan's Tadami project', *Water Power & Dam Construction*, August, pp. 33–39

1986(5) FORD, C.R., and ELDRIDGE, M.D.: 'Planning, design and execution of movement of naval test plant from Barrow to the north coast of Scotland', *Proc. ICE*, Part 1, Aug., pp. 885–902

1986(6) MEGNINT L.: 'The new design of tidal bulb units based on the experience from the Rance tidal station and on-the-river bulb units'. Proc. 3rd International Symposium on Wave, Tidal, OTEC and Small Scale Hydro Energy, Brighton, England, May, BHRA

1986(7) HILLAIRET, P., and WEINSTOCK, G.: 'Optimising production from the Rance Tidal Power Station'. *Ibid*

1986(8) SHARMA, H.R.: 'Kachchh tidal power project'. Proc. 3rd International Symposium on Wave, Tidal, OTEC and Small Scale Hydro Energy, Brighton, England, May. BHRA

1986(9) SEVERN TIDAL POWER GROUP: 'Tidal power from the Severn'.

1986(10) PAUL, T.C., and DILLON, G.S.: 'Dimensioning vertical lift gates', *Water Power & Dam Construction*, Nov., pp. 45–47

1986(11) YULIN, H.E., and TSENG HUANG: 'Empirical formulae for gate weights', *ibid.*, pp. 52–55

1986(12) HYDRAULICS RESEARCH Ltd. (Wallingford).: 'An analysis of the behaviour of fluid mud in estuaries'. Report SR 84, March

1986(13) KOREAN OCEAN RESEARCH DEVELOPMENT INSTITUTE: 'Report on studies of Garolim tidal power project. Vol II'. Technical report. Seoul, S. Korea

1987(1) House of Commons debate on: Energy (Renewable Sources). *Hansard*, 30 Oct., pp 564–624

1987(2) SCHWEIGER, F., and GREGORI, J.: 'Developments in the design of Kaplan turbines', *Water Power & Dam Construction*, Nov.

1987(3) ANON: 'Brazil plans tidal power plant', *Water Power & Dam Construction*, Oct.

1987(4) DELORY, R.P.: 'Prototype tidal power plant achieves 99% availability', *Sulzer Technical Review*, Jan

1987(5) ANON: 'New threat to Mersey', *Bird Watching*, June

1987(6) BESSIERE, C.: 'Power generation studies of a double reservoir scheme along the Constantin coast in France', *Tidal Power*, pp. 149–152

1987(7) 'Renewable Energy in the UK: The way forward'. Energy Paper 55, HMSO

1987(8) BINNIE & PARTNERS for Dept. of Energy: 'Models for tidal power'. Jan

1987(9) YARD LTD.: 'Air turbines for power generation on tidal barrages'. March

1987(10) BINNIE & PARTNERS for Dept. of Energy: 'Turbines for small scale tidal power'. April

1987(11) COLLINS, M.: 'Sediment transport in the Bristol Channel: A review', *Proc. Geol. Ass.* 98(4) pp. 367–383

1987(12) GILFILLAN, W.A., MACKIE, G.C., and ROWAN, R.P.: 'Steel caissons for tidal barrages', *Tidal Power* (Thomas Telford), pp 331–330

1987(13) BAKER, A.C.: 'The development of functions relating cost and performance of tidal power schemes and their application to small scale sites', *Tidal Power* (Thomas Telford), pp. 331–344

1987(14) KEILLER, D.C.: 'Some interactions between tides and tidal power schemes in the Severn estuary', *Tidal Power*, pp. 17–28

1987(15) PUGH, D.T.: 'Tides, surges and mean sea-level' (John Wiley) pp. 60–85

1987(16) KERR, D., MURRAY, W.T., and SEVERN, B.: Civil engineering aspects of the Cardiff–Weston Barrage', *Tidal Power* (Thomas Telford), pp. 41–70

1988(1) ADMIRALTY tide tables and tidal stream tables, Vols. 1, 2, 3. Published annually by the Hydrographer to the Navy. (HMSO)

1988(2) HOCKIN, D.C., and PARKER, D.M.: 'The effects of development of a tidal barrage upon the water and sediment quality of the Mersey estuary (UK) and its biota', *Wat. Sci. Tech.*, 20 (6/7), pp. 229–233

1988(3) McKINLEY, R.S., PATRICK, P.H., and MUSSALI, Y.: 'Controlling fish movement with sonic devices', *Water Power & Construction*, March, pp. 13–17

1988(4) STERN, M., and ROWE, J.: 'A new perspective on the Severn Barrage'. M. Stern, Feb.

1988(5) RENDEL, PALMER & TRITTON for Dept. of Energy: 'Sand embankments and diaphragm walling for tidal power barrage construction'. March

1988(6) SOLOMON, D.J.: for Dept. of Energy: 'Fish passage through tidal energy barrages. Final report', 1988

1988(7) MERSEY BARRAGE COMPANY: 'Tidal power from the river Mersey: A feasibility study: Stage I report'.

1988(8) HYDRAULICS RESEARCH LTD.: 'Estuarine muds manual'. Feb.

1988(9) HYDRAULICS RESEARCH LTD.: 'The behaviour of estuarine muds during tidal cycles: An experimental study and mathematical model'. Feb.

1988(10) HYDRAULICS RESEARCH LTD.: 'Laboratory investigation of measures to reduce mud siltation in dredged navigation channels'. Report SR 162, Feb.

1988(11) BINNIE & PARTNERS for Llanelli Borough Council: 'River Loughor Tidal Power Scheme: Report on the feasibility study'. Nov.

1988(12) HYDRAULICS RESEARCH LTD.: 'A two-dimensional model of the movement of fluid mud in a high energy, turbid estuary'. Report SR 147, Jan

1989(1) BINNIE & PARTNERS for Dept. of Energy: 'The UK potential for tidal energy from small estuaries; a survey of possible small sites on the west coast'. Jan

1989(2) STEEL CONSTRUCTION INSTITUTE for Dept. of Energy: 'Generic steel designs for tidal power barrages'.

1989(3) BINNIE & PARTNERS for the Humber Barrage Group: 'Humber tidal barrage pre-feasibility study'.

Index